ABOUT THE AUTHOR

Christopher Barnatt has been a professional futurist for over 20 years, and runs ExplainingTheFuture.com, Explaining Computers.com and their associated YouTube channels. He has published ten previous books and numerous articles, with over 200 contributions to broadcast, print and online media around the globe.

For 25 years Christopher lectured in computing and future studies in Nottingham University Business School, where he spent seven years as Director of Undergraduate Programmes. As a keynote speaker, he now delivers presentations for a wide range of clients in sectors including financial services, healthcare and the arts. Since the 1980s, Christopher has also worked as an animator, with footage supplied to the BBC and other broadcasters. You can follow him at twitter.com/chrisbarnatt.

By the same author:

BOOKS

3D Printing: Second Edition
3D Printing: The Next Industrial Revolution
Seven Ways to Fix the World
25 Things You Need to Know About the Future
A Brief Guide to Cloud Computing
Cyber Business: Mindsets for a Wired Age
Challenging Reality: In Search of the Future Organization
Valueware: Technology, Humanity and Organization
Management Strategy & Information Technology
The Computers in Business Blueprint

YOUTUBE CHANNELS

YouTube.com/ExplainingTheFuture
YouTube.com/ExplainingComputers

WEBSITES

ExplainingTheFuture.com
ExplainingComputers.com

THE NEXT BIG THING

From 3D Printing to Mining the Moon

Christopher Barnatt

ExplainingTheFuture.com

First published by ExplainingTheFuture.com®

For press, rights, translation and other enquiries,
please e-mail chris@explainingthefuture.com

Copyright © Christopher Barnatt 2015.

The right of Christopher Barnatt to be identified as the author of this work has been asserted by him in accordance with the Copyright, Designs and Patents Act 1988.

All rights reserved. This book is sold subject to the condition that it shall not, by way of trade or otherwise, be lent, re-sold, hired out or otherwise circulated in any form of binding or cover other than that in which it is published and without a similar condition including this condition being imposed on the subsequent purchaser.

ISBN-10 : 1518749577

ISBN-13 : 978-1518749575

Printed and bound on demand.

Typeset in Adobe InDesign by Christopher Barnatt.

Disclaimer

While every effort has been made to ensure that the content in this book is as accurate as possible, no warranty or fitness is implied. All trademarks included in this book are appropriately capitalized and no attempt is made or implied to supersede the rights of their respective owners.

1 3 5 7 9 10 8 6 4 2

To my former colleagues in
Nottingham University
Business School

CONTENTS

Acknowledgements	ix
Prologue: Beyond the Internet	1
Part I: Local Digital Manufacturing	
1. 3D Printing	15
2. Synthetic Biology	43
3. Nanotechnology 2.0	69
Part II: Synthetic Citizens	
4. Artificial Intelligence	101
5. Humanoid Robots	133
Part III: Resources from Space	
6. Space-Based Solar Power	161
7. Asteroid Mining	187
8. Mining the Moon	213
Part IV: Transhuman Evolution	
9. Post-Genomic Medicine	243
10. Cyborg Synthesis	270
Epilogue: Toward the Singularity	297
References	306
Index	318

ACKNOWLEDGEMENTS

This book has been written during a period of personal uncertainty and change. The actual research and writing process was intense but uneventful. Yet while I have been writing about the world of tomorrow, I have had to make some big decisions concerning my own future, and this has not made the production of *The Next Big Thing* as straightforward as it could have been.

The good news is that, over the past year, I have received a great deal of excellent advice and support from many people. In particular, I would like to thank my parents for their help in my lengthy decision making process. My thanks also to Mark Daintree, Ken Starkey, Sue Tempest, Stephen Diacon, Teresa Bee, Wendy Chapple, Tracey Bettinson, Steve Moore, George Kuk and David Newton. All of you have been involved in many very long conversations that have resulted in my embarkation on, as a wise man recently told me, 'my next great adventure'.

For her meticulous proofreading and other support, I would like to thank Kathleen Visser. Thanks also to Mark Daintree, Sally Hopkinson, Victoria Wrigley and Rozina Shaikh for their input to the cover, which fortunately proved far less problematic to finalize than last time. I am glad we got the Moon sorted out! Thanks also to Agi Haines for her contribution to chapter 10.

And lastly, a nod of appreciation to my YouTube subscribers and other 'followers' in social media. The information and ideas included in this book – and which I will incorporate into many future videos and other media – will hopefully keep us all engaged in a lively discussion for some time . . .

Christopher Barnatt,
November 2015.

PROLOGUE
BEYOND THE INTERNET

In 1995 I published a book called *Cyber Business* that predicted the birth of e-commerce and social media. Since that time I have frequently been quizzed about future online trends, and have introduced many individuals and organizations to key developments like cloud computing. Even so, when people now ask me about the world of tomorrow, I often begin by stating that the Internet Revolution has come to an end.

The mainstreaming of the Internet in the closing years of the 20th century was undoubtedly a seminal moment in human history. Nevertheless, no period of revolution can go on indefinitely. Look back 375 million years, and our ancestors were dragging themselves out of the oceans and developing a technology called 'lungs' in what we could term the 'Breathing Air Revolution'. To this day, lungs remain critical to the survival of all mammals. And yet the Breathing Air Revolution has clearly long since ended.

In a similar fashion, now that the Internet has become established as our collective, planetary nervous system, so the idea of the 'Internet Revolution' ought rapidly to be consigned to history. Granted, the online world will continue to incrementally evolve. But those seeking radical, future shaping innovation really need to divert their attention away from the cyber world to which we increasingly retreat.

While our civilization is now reliant on the Internet, we are even more dependent on the sustainable production of physical things. We therefore need to shift our collective focus away from the digital world, and toward the innovation of radical new manufacturing methods and the attainment of fresh resource supplies. If only to ensure our civilization's long-term survival, it is now time for the human race – and the world's stock markets – to recover from Internet fever in preparation for the Next Big Thing.

The following chapters extrapolate from bleeding-edge science and engineering to predict ten dominant technologies and related undertakings of the 2020s, 2030s, 2040s and beyond. In doing so, the four parts of this book additionally highlight four fundamental future transitions. These will take us on a journey from the reign of the microprocessor to that of the microfabricator; from a use of dumb computing devices to a cohabitation with smart synthetic citizens; from consuming less here on the Earth to finding more resources out in space; and from healthcare systems focused on medical maintenance to those which champion generational upgrades of the human form.

While each of the ten Next Big Things that will underpin the above transitions could develop and be studied in isolation, it is my contention that they are all highly interrelated. Understanding these interrelations is also quite important. So, before we get to chapter 1 – and to provide some explanation of the last paragraph! – here is a brief overview of things to come.

LOCAL DIGITAL MANUFACTURING

The current model of industrial production works roughly like this. Somebody dreams up a new product, a factory thousands of miles from most potential customers is tooled up to produce it, and a large number of identical products are manufactured in the hope that somebody, someday will want to buy them. The products are then transported to a warehouse, from where they are gradually shipped to wholesalers and retailers for potential sale.

In time, many products are sold, although about one-seventh of the resultant revenues are spent on transportation, warehousing and related logistics services. Unfortunately, some of the products that are manufactured are never actually bought by anybody, and need to be discarded. This highly wasteful, globalized, mass production arrangement works because there are enough relatively cheap natural resources still available for companies to be able to squander a great deal and still make a profit. But as resources become less plentiful and energy prices rise, so how most things are made will have to change. And this is when we will transition to local digital manufacturing.

As detailed across Part I of this book, local digital manufacturing (LDM) uses digital technologies to make products on demand very close to where final consumers actually live. Using future LDM hardware, designs for both inorganic and organic products will be able to be stored and transported digitally, before being locally 'materialized' into a physical format one layer, cell or molecule at a time.

Right now, as we shall explore in chapter 1, the most developed LDM technology is 3D printing. This builds objects in layers, and even today can fabricate items in plastics, metals, ceramics, foodstuffs and living cells. Already jewelry, car bodies, toys, aerospace components, medical devices, works of art and buildings have been 3D printed. Using an organic 3D printing variant called 'bioprinting', living human tissue has also already been manufactured one layer at a time. Of all the Next Big Things detailed in this book, 3D printing may well be the first to enter the mainstream. Though when it comes to local digital manufacturing, 3D printing will be just the tip of the iceberg.

In addition to 3D printing, there is already another highly versatile technology that can turn digital designs into complex physical things in pretty much any location. This amazing technology is life itself, with the DNA of all plants and animals containing a robust digital code that can tell cells how to reproduce, rearrange and subsequently function. So what if we could turn living biology into a construction kit that could be digitally programmed as a production technology? Well, we have already started to do this with the creation of a new science called synthetic biology.

As we shall see in chapter 2, synthetic biology allows living things to be created that have never existed in nature. Already synthetic biology is being applied to develop microorganisms that can ferment organic feedstocks into biofuels, bioplastics, bioacrylics and pharmaceuticals. In time, synthetic biology even has the potential to create new plants and novel animals for specific manufacturing purposes. Consumables for 3D printers, for example, may one day be grown locally in desktop hydroponic devices or urban vertical farms.

As well as relying on 3D printing and synthetic biology, LDM will be facilitated by next generation nanotechnologies. As we will investigate in chapter 4, so-termed 'atomically precise manufacturing' (APM) will permit objects to be fabricated on a molecular scale using a process called 'self-assembly'. Over the next two decades, we are also likely to witness the convergence of nanotechnology with 3D printing and synthetic biology. In turn, this will facilitate the construction of 'microfabricators' that will be able to fashion a very wide range of highly sophisticated products directly from digital designs.

Even today there is an overlap between 3D printing, synthetic biology and nanotechnology, with scientists and engineers in each discipline increasingly sharing knowledge and techniques as they learn to digitally manipulate matter on a very small scale. For example, nanoscale 3D printing processes are starting to be developed that can allow material composition as well as material placement to be digitally controlled. Add in synthetic biology, and future microfabricators should be able to control the composition, placement and living behaviour of digitally manufactured things. As I said a few paragraphs back, 3D printing will be just the start of the local digital manufacturing revolution.

SYNTHETIC CITIZENS

A score or less years hence, we are very likely to be sharing our first planet with artificial entities more intelligent than ourselves. As we will discover in chapter 4, some of these will be disembodied artificial intelligences (AIs) that will help human beings with specific tasks at which machines tend to excel. Already so-termed 'narrow' forms of AI are able to pilot aeroplanes, drive automobiles, diagnose disease, manage power grids, track vehicle license plates, translate languages, and perform stock market trades. Many people are also starting to use 'virtual assistants' (VAs) like Microsoft's Cortana or Apple's Siri, with the trend to develop AI as a next-generation computing interface set to continue.

Today, Cortana or Siri are novel add-ons bundled with an operating system. Yet in less than 10 years, Microsoft or Apple's primary product may well be a virtual assistant with an operating system and supportive hardware bundled on top. This means that, sometime in the 2020s, we may talk far more about VAs and far less about PCs. Indeed, if you are wondering why there is not a section of this book devoted to future computing, it is because I suspect that we will soon look back on the use of dumb computing devices as a rather quaint late-20th and early-21st century phenomenon.

Exactly when and how 'artificial general intelligences' (AGIs) will be created is a point of significant contention. There are also many who believe that creating highly sophisticated AGIs is a dangerous undertaking that ought to be prevented, or at least very tightly controlled. Personally, I think that the development of AGIs is not just inevitable, but essential if we are to rollout widespread local digital manufacturing and in the process to deal with looming resource scarcity. Granted, current legal frameworks will need to be adapted to deal with very smart technology that can act autonomously and potentially do both very good and very bad things. We may even choose to give future AGIs some legal rights. Nevertheless, the real debate ahead will, I think, be far more about the role to be played in our society by non-human forms of intelligence, rather than whether or not they should be created.

In practical terms, it is likely to be the mainstream rollout of autonomous vehicles in the 2020s, and humanoid robots in the 2030s and 2040s, that will bring the critical debates surrounding AI to the fore. Within 15 years, most people will either be travelling in driverless vehicles, or will be relying on the autonomous carriages occupied by others not to crash into their car or to run them over on the sidewalk. As we shall explore in chapter 5, in a few decades time it is also very likely that humanoid robots will be delivering healthcare, looking after the elderly, and transforming at least some traditional production methods. Our artificial world has been crafted for occupation and manipulation by the human form. It will therefore make sense to build mechanical beings in our own image, even if doing so may not be to everybody's taste.

Robot ascendance is likely to be symbiotically associated with the rise of local digital manufacturing. On the one hand, robots will become a complimentary technology to 3D printing and synthetic biology, as they will be able to locally prepare and assemble product parts and raw materials that are fabricated on demand. The other way around, local digital manufacturing technologies will be critical in robot evolution, as they will become the dominant means of robot procreation. Even today, robot development is being driven forward not just by the availability of cheap computing power, but due to the existence of low-cost 3D printers that can rapidly and cost-effectively produce custom mechanical components.

As synthetic biology and organic computing develop, it is also quite likely that parts for future robots will be able to be grown. Today, the popular vision of a humanoid robot is of a metal or plastic machine. But in three decades time, our mechanical servants and companions are just as likely to be constructed from living tissue, or else from materials produced via a synthetic organic process. Who knows, 20 or 30 years from now, you may have a warm bucket in a kitchen cupboard in which you are growing a new arm for your favourite android companion.

RESOURCES FROM SPACE

Today, a great deal of attention is starting to be focused on conducting our lives and operating our businesses in a 'sustainable' fashion. Usually this involves attempts to use fewer resources and to reduce our carbon footprint. Most of the time, such initiatives are a great idea. They will also be boosted by innovations in local digital manufacturing that will enable people to produce things using less energy and fewer raw materials.

The above points noted, all current and future attempts to become 'sustainable' can at best constitute a short-term solution to the resource requirements of future generations. Like it or not, it is a physical certainty that the raw materials and energy sources available on the Earth are finite. This means that, in the long-term, the survival of our civilization has to depend on obtaining fresh energy and raw material supplies from beyond our first planet. At least some of the AIs and robots referred to in the previous section will therefore spend their lives obtaining resources from space.

Across Part III of this book we will investigate a wide range of possibilities for extraterrestrial power generation and off-world mining. Staying closest to the Earth, chapter 6 will first detail how 'space-based solar power' (SBSP) could be developed. Future SBSP systems would place solar power satellites in geosynchronous orbit. These would then beam energy to 'rectennas' on the Earth using microwaves or lasers.

NASA began feasibility studies into SBSP in the 1970s. More recently, in April 2014 the Japan Aerospace Exploration Agency (JAXA) revealed a roadmap for a SBSP system to provide energy to Tokyo in the 2030s. The creation of such a system will require significant improvements in all aspects of space technology, as well as the potential development of entirely new means for getting into orbit. In addition to providing an overview of SBSP possibilities, chapter 6 will therefore examine the feasibility of future 'space elevators' that may help off-world energy production to become a reality.

While building solar power satellites may help to meet some of our future energy needs, it will not assist with the supply of physical resources. In chapter 7, we will therefore consider the possibility of mining the asteroids – a proposition already being taken very seriously by two foresighted companies called Planetary Resources and Deep Space Industries. I would already place a fairly safe bet that many of today's young people will one day own a consumer product manufactured at least in part from asteroid deposits.

In addition to SBSP and asteroid mining, we are at some point also likely to return to the Moon in search of new energy and raw material supplies. A potentially very valuable future nuclear fuel called helium-3 is relatively abundant in the lunar regolith, while our lonely satellite is also thought to harbour substantial deposits of cobalt, iron, gold, palladium, platinum, titanium, tungsten and uranium.

Since 2009, NASA experiments have also confirmed the presence of water on the Moon. This could prove critical in supporting long-term human occupation, as well as providing a source of oxygen and hydrogen for rocket fuel. Chapter 8 will examine a range of options for lunar resource utilization, including the potential development of lunar space elevators and large-scale, Moon-based 3D printers.

TRANSHUMAN EVOLUTION

By the second half of this century, a growing proportion of the world's population will be a mashup of legacy biology and artificial digital technologies. Such 'transhumans' will have had their bodies or brains augmented using technologies including bioprinting, synthetic biology, nanotechnology, cybernetics and genetic medicine. The latter is the subject of chapter 9, where we will examine how – in a new age of post-genomic healthcare – doctors and AI systems are set to become programmers of human DNA. The 21st century is likely to be remembered as the historical period in which humanity took conscious control of its own evolution, and when the line between 'natural' creation and 'artificial' technology became irrevocably blurred.

As we shall explore in chapter 10, a cybernetic synthesis of human beings and machines is an almost inevitable consequence of the development of local digital manufacturing, the creation of robots and AI, and the pursuit of resources from space. Why? Well, for a start, as we learn to digitally manufacture products one layer, cell or molecule at a time, so we will also hone the skills necessary to take digital control of human biology. Since November 2014, bioprinting pioneer Organovo has been 3D printing human liver tissue as a commercial product (if currently for drug testing purposes), while the line between synthetic biology and genetic medicine is already tantalizingly thin.

In the short-term, legal and ethical constraints on the manipulation and adaptation of the human body may limit the extent to which future scientists and engineers will be able to 'play god'. But given that local digital manufacturing will empower individuals as well as corporations to inorganically and organically fabricate anytime, anyplace and anywhere, it seems inconceivable that its technologies will not be widely applied both in healthcare and to facilitate future human augmentation.

Some people may question why, later this century, anybody would want to merge with artificial technology. In part the answer is simply that the pursuit of excellence remains a common individual goal and the driving force of our evolution. More pragmatically, as AIs and humanoid robots become both more intelligent and more physically dexterous than human beings, so it is very probable that at least some people will want to 'keep up with the machines'.

Bioprinted or synthetically grown components for future humanoid robots may well be biocompatible with a genetically re-engineered human anatomy. So when somebody sees a robot strolling down the road with a cool pair of legs, they could favourite the design and have it downloaded and replicated for themselves. Some people in the future may even swap body parts as regularly as we currently change hairstyles or clothes. It may even become common to exchange limbs, eyes, memory circuits and information processing hardware with robotic co-workers or friends.

The above factors noted, the biggest driver of the most extreme form of cyborg synthesis is going to be our requirement to obtain resources from space. To achieve this goal, our civilization will need to send highly adaptable and intelligent beings far from the Earth, and not all of these will be able to be entirely robotic.

Unfortunately, the current human form is about as well equipped to live in space as a fish is suited to reside on dry land. Yes, humans can protect themselves in space suits and pressurized capsules, and can shelter behind radiation shielding when required. We can also take food, water and oxygen with us from the Earth – or learn to obtain such critical life support supplies off-world. Although, when it comes to large-scale space endeavours, it is going to prove far safer and far more cost effective for future space pioneers to be transhumans with bodies designed for long-term deep-space survival. By the end of this century, we are therefore likely to witness the emergence of a new cybernetic superspecies who will not be reliant on oxygen, water and food, and who will be far less easily damaged by extraterrestrial radiation than their traditional human forebears.

TOWARD THE SINGULARITY

In 1961, science fiction author and futurist Arthur C. Clarke wrote that 'any sufficiently advanced technology is indistinguishable from magic'. Certainly, a great many of the technologies that we currently take for granted would have seemed magical only a century ago. Yet many of the innovations on the medium- and long-term horizon are destined to be even more astonishing. Microfabricators, synthetic citizens, resources from space, and our transhuman evolution, may therefore be perceived as impossible fantasies by the majority of the world's population.

Hopefully, as a reader of this book, you are distinct from the majority and more open than most people to accept the incredible developments and opportunities that lie ahead. As I have detailed in this *Prologue*, four fundamental transitions now loom on the horizon, and will soon drive radical changes in how things are made, who we share the planet with, where resources come from, and the evolution of the human race. Each of these transitions is going to be an extraordinary adventure. I am therefore pleased that, by choosing to read *The Next Big Thing*, you have decided to proactively step on board to anticipate the ride.

In aggregate, all of the technologies and undertakings explored in this book will lead us toward a moment in history called the 'Singularity'. This is a technological event horizon beyond which we cannot see, and that we will reach when exponential progress makes possible anything we can imagine.

Looked at from one perspective, we will arrive at the Singularity when the divide between 'technology' and 'magic' blurs. Or to reduce things to a more practical level, the Singularity will be upon us when we are able to digitally program, replicate, repair and otherwise control all forms of living or inorganic matter. At the Singularity and beyond, we will also no longer face any resource constraints, as we will have learned how to turn waste products into fresh raw materials, or to access the very broad range of resources waiting for us beyond Planet Earth.

Journeying toward the Singularity is likely to require the application of mental and physical capabilities far beyond those of the current human form. The creation of artificial intelligences and very sophisticated robots will therefore be essential if we are to arrive in a new age of enlightenment and plenty. Also very likely to be required will be some transhuman upgrading of *Homo sapiens*.

However magical it may sound, the Singularity is a point in future history that many of today's young people will one day experience, and which could turn out to be the greatest ever Next Big Thing. We live in extraordinary times that are going to get increasingly unbelievable. So, without further introduction, it is now time for us to explore the incredible possibilities that lie ahead . . .

PART I

LOCAL DIGITAL MANUFACTURING

1
3D PRINTING

The human race has become a rather wasteful species. Billions of slightly-broken products are discarded every year, while poor market forecasts and overproduction result in manufactured items that never get sold. Most products also travel thousands of miles to reach their final customer. In fact, about one seventh of the energy reserves and raw materials consumed by humanity are devoted to transporting things around the planet. Economists may preach that this is a fantastic way to sustain an increasingly global economy. Yet, in reality, it is an intensely foolish and potentially suicidal state of affairs.

To help our civilization to survive and thrive, it would be far better to locally manufacture the majority of products on-demand, and to repair rather than discard those that fail. It is therefore fortunate that a wide range of innovations in local digital manufacturing (LDM) that will allow this to happen are already being honed. As we shall see in chapter 2 and chapter 3, critical LDM technologies include synthetic biology and next generation nanotechnology. But before these enter mainstream application, the dominant form of LDM will be 3D printing.

3D printing is the popular term for 'additive manufacturing', and refers to all technologies that turn digital models into physical objects by building them up in layers. In effect,

3D printing does exactly what it says on the tin. Just as traditional 2D printers create documents or photographs by outputting one or a few layers of ink, so 3D printers create real, solid, fully-dimensional stuff by outputting hundreds or thousands of layers of a build material.

As we shall see in this chapter, 3D printing has already been used to fabricate a wide range of things that include product prototypes, prosthetics, rocket engine components, and a five storey apartment block. The 3D printing industry is also growing rapidly, with Deloitte estimating the sale of 220,000 3D printers in 2015, and Gartner predicting 496,000 units to be sold in 2016. According to a June 2015 report published by Smithers Pira, the 3D printing marketplace will be worth $49.1 billion by 2025.

In the future, 3D printing has the potential to do for physical objects – or 'physibles' – what the Internet has already done for the digital creation, storage, manipulation and communication of information. What this means is that 3D printing could allow products to be designed, manipulated, stored and transported in a digital format, before being 'materialized' on demand anytime, anyplace and anywhere. The potential implications for global manufacturing, the logistics industry and personal fabrication are subsequently breathtaking.

3D printing has, in fact, already been hailed as the technology that will destroy capitalism by putting the means of production into the hands of the majority. This, I am certain, is too extreme a view. Nevertheless, with 3D printing rapidly evolving from a prototyping process into a mainstream production technology, it is easy to see why many believe that 3D printing will be the Next Big Thing.

MATERIAL EXTRUSION

Contrary to popular belief, 3D printing is not a single technology. Rather, there are already a wide range of methods for

Material Extrusion:	a nozzle extrudes a semi-liquid material to build up successive object layers.
Vat Photopolymerization:	a laser or other light source solidifies successive object layers on the surface or base of a vat of liquid photopolymer.
Material Jetting:	a print head sprays a liquid that is either set solid with UV light, or which solidifies on contact.
Binder Jetting:	a print head selectively sprays a binder onto successive layers of powder.
Powder Bed Fusion:	a laser or other heat source selectively fuses successive layers of powder.
Directed Energy Deposition:	a laser or other heat source fuses a powdered build material as it is being deposited.
Sheet Lamination:	sheets of cut paper, plastic or metal are stuck together.

Figure 1.1: 3D Printing Technologies. Based on ASTM F2792.

turning a digital design into a physical object by building it up in layers. In an attempt to introduce some clarity into a quite confusing marketplace, in June 2012 the American Society for Testing and Materials (ASTM) categorized all 3D printing technologies under seven generic headings. These are outlined in figure 1.1, and provide a useful taxonomy for understanding current and potential future 3D printing methods.

Right now, the most common form of 3D printing is material extrusion. This was invented by Scott Crump in 1998 following a successful attempt to build a plastic frog for his daughter using a hot glue gun. Crump subsequently

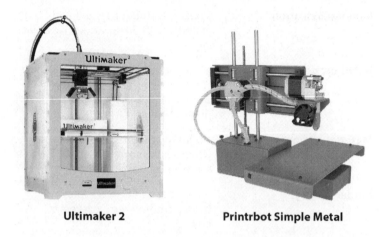

Ultimaker 2   **Printrbot Simple Metal**

Figure 1.2: Desktop 3D Printers from Ultimaker & Printrbot.
Images courtesy of respective manufacturers.

founded 3D printing giant Stratasys, and patented his material extrusion process under the name 'fused deposition modelling' (FDM).

In a typical material extrusion 3D printer, a thermoplastic build material known as 'filament' is fed to a print head where it is heated into a molten state and extruded onto a 'build platform' to form an object layer. The build platform then lowers a fraction, another layer of thermoplastic is extruded, and so on.

3D printers that extrude thermoplastics come in a wide variety of sizes, prices and configurations. At the time of writing, the cheapest desktop models start at about $300, with prices likely to fall to $99 well before 2020. At the other end of the spectrum, industrial '3D production systems' cost up to $900,000. Figure 1.2 illustrates two popular desktop 3D printers from Ultimaker and Printrbot, while figure 1.3 shows four Stratasys 3D production systems in a factory environment.

3D PRINTING

Figure 1.3: Stratasys Fortus 900mc 3D Production Systems.
Photo courtesy of Stratasys.

Although the cheapest material extrusion 3D printers can currently only build objects in a single thermoplastic, printers with dual or triple extruders that can output multiple colours of the same thermoplastic are becoming quite common. 3D printers with 'mixer extruders' that can combine several different filaments in their print head – so permitting full-colour plastic printout – have also been demonstrated. It is therefore not unreasonable to assume that, in five or ten years time, full-colour material extrusion 3D printing will be a mainstream technology.

The most common build materials used for material extrusion are the petroleum-based thermoplastics acrylonitrile butadiene styrene (ABS), nylon and polycarbonate (PC), as well as a bioplastic called polylactic acid (PLA). 'Thermoplastic elastomers' are now also available that permit the material extrusion of rubber-like, flexible parts.

In the past few years, a range of composite 3D printing filaments have additionally arrived on the market. These include thermoplastics mixed with wood, carbon fiber, brick, or various metals. For example, a thermoplastic composite called copperFill was launched by a filament manufacturer called Colorfabb in 2014. This contains a fine copper powder, and allows consumer 3D printers to make metal-like objects that can be polished to a shine. Pioneer Graphene 3D Lab is even working on nanocomposite filaments that combine a thermoplastic with carbon nanotubes or graphene, and which it hopes will soon allow the 3D printout of working electronic devices. As all of these developments illustrate, by the 2020s it is highly probable that the $99 hardware I predicted a few paragraphs back will be able to print in a wide range of materials.

Indicating even broader future possibilities, several companies are developing material extrusion 3D printers that output thermoplastics reinforced with another material during printout. Most notably, MarkForged in the United States now market a 3D printer that combines a thermoplastic with a continuous strand of carbon fiber, fiberglass or Kevlar. This allows plastic objects to be 3D printed that are stronger than some metals.

In addition to thermoplastics, or composites thereof, it is already possible to 3D print by extruding metals, concrete, clay or food. When it comes to extruding metals, some of the leading work is being done at the University of Cranfield for aircraft manufacturer BAE Systems. Here a 3D printing method called 'wire and arc additive manufacturing' (WAAM) has been created that melts a titanium strand fed to a computer-controlled arm. Already WAAM technology has been used to create a 1.2 metre spar section of an aircraft wing. Eventually it is hoped that entire aircraft frames will be 3D printed.

Also 3D printing in metal using a material extrusion process are Sciaky. Here the involved technology is termed electron beam additive manufacturing (EBAM), and feeds two solid metal wire feedstocks into an electron beam that fuses them into large, industrial metal parts. The resultant printouts have clearly stepped layers, but may be post-processed by CNC machining to achieve a smooth surface. Build materials for EBAM currently include titanium alloys and tantalum.

Since 2004, Professor Behrokh Khoshnevis at the University of Southern California has been working to develop 3D printers capable of extruding concrete. This he labels 'contour crafting', which he in turn describes as a 'mega scale layered fabrication process' for use on building sites. Similarly working to 3D print human dwellings are the World's Advanced Savings Project (WASP), who are based in Italy. Their experimental 'Big Delta' 3D printer extrudes clay or soil mixed with resin, with the intention being to develop the technology to rapidly 3D print houses or modular building sections following natural disasters.

On a less grand scale, several pioneers have managed to 3D print many different kinds of food, including chocolate, ice-cream, candy and cake frosting. My favourite team is a Spanish research group called Robots in Gastronomy, who have developed a 3D printer called FoodForm. This can extrude food onto any surface, including a hot grill or a chilled anti-griddle. Using the FoodForm, Robots in Gastronomy have already managed to 3D print by extruding bread and cookie doughs, chocolate creams, cheese, ice cream, cheesecake and various frostings.

SOLIDIFYING LIQUIDS

Alongside material extrusion, two other common 3D printing technologies are vat photopolymerization and material jetting. These are currently more accurate than material extrusion, offer fantastic surface quality, and use liquid build materials.

Vat photopolymerization was invented by Charles 'Chuck' Hull in 1983, and uses a laser beam to trace out object layers on the surface or base of a vat of liquid photopolymer resin. Where it is contacted by the laser, the photopolymer cures from liquid to solid. When each layer is complete, the 3D printer's build platform then moves a fraction, so allowing the next layer to be traced out and set solid. Chuck called this 3D printing process 'stereolithography', and his 3D printer a 'StereoLithographic Apparatus' (SLA). In 1986 he also obtained patent protection for his revolutionary innovations, and started a company called 3D Systems. Three decades on, 3D Systems is still going strong, and alongside Stratasys is well established as one of the two giants of the 3D printing industry.

Material jetting 3D printers spray liquid photopolymer layers from an inkjet-style, multi-nozzle print head. Each layer is then set solid with UV light before the next layer is added. Material jetting hardware has been developed by many companies including Stratasys (who term it 'PolyJet') and 3D Systems (who call it 'MultiJet Printing'). As illustrated in figure 1.4, the Objet Connex hardware from Stratasys is particularly impressive, as it can 3D print multi-material, multi-colour objects by supplying different photopolymers to the print head which are then combined as required.

ADHERING POWDERS

Our next three 3D printing technologies – binder jetting, powder bed fusion and directed energy deposition – all turn a digital file into a solid object by sticking together granules of a fine powder. In binder jetting hardware, successive layers of powder are rolled or otherwise swept across the build area. A multi-nozzle print head then sprays on a binder solution to stick the required powder granules together. In some binder jetting printers, the print head also deposits coloured inks, so allowing full-colour objects to be created.

For many years binder jetting was almost entirely limited to making objects in a gypsum-based powder. This then had

Figure 1.4: A Stratasys Objet500 Connex3.
Photo courtesy of Stratasys.

to be infiltrated with resin after printout if a robust object was required. However, in late 2013, 3D Systems launched a printer – the ProJet 4500 – that can 3D print full-colour, semi-rigid plastic parts that require no post-processing after their removal from the print bed. German 3D printer manufacturer voxeljet additionally now manufacture binder jetting 3D printers than can build objects from a modified acrylic glass powder, or from sand to produce casting molds. Some of the binder jetting printers manufactured by voxeljet are also very large indeed. Most notably, their VX4000 can 3D print objects as large as 4 x 2 x 1 metres in size.

A few binder jetting 3D printers, such as those manufactured by a company called ExOne, can create objects out of powdered metals, including bronze and the nickel-based alloy Inconel 625. After metal parts are 3D printed using ExOne's technology, they need to be cured in an oven, and placed in a kiln for 24 hours to be infused with more metal. The result is a 3D printed object that is about 99.9 per cent solid.

To 3D print really high quality metal parts, an alternative technology called powder bed fusion has to be used. This is similar to binder jetting, save that here the layers of powder are selectively fused solid using a controlled heat source. Most commonly this is a laser, with the involved process termed laser sintering or laser melting. Another variant of powder bed fusion is electron beam melting (EBM), which uses an electron beam to fuse metal powder granules in a vacuum.

In addition to allowing the 3D printout of metal objects, powder bed fusion is commonly used to produce high quality parts in plastics (such as nylon), or plastic-metal composites. Powder bed fusion can, for example, 3D print using a material called alumide, which is a mix of nylon and aluminium powders.

A final, powder-based 3D printing technology is directed energy deposition (DED), also known as 'laser powder forming' or 'laser engineered net shaping' (LENS). Here, a metal powder is supplied to a print head that moves in 3D space, and which jets the powder into a high-power laser beam. Because the metal powder is not laid flat for selective fusion, directed energy deposition can either create new objects, or fuse new metal onto existing parts. Most notably, the technology is already being used to repair worn or otherwise damaged jet engine turbine blades.

Directed energy deposition can fabricate objects in new alloys by feeding a range of different metal powders to the print head. The powders can even be mixed in continuously variable ratios during printout. This presents the potential to 3D print components in novel metal alloys, and indeed out of alloys that have different properties in different parts of an object.

BINDING SHEETS

The last, current 3D printing technology is sheet lamination. This builds objects by sticking together layers of cut paper, plastic or metal foil. Where objects are built out of cut paper – as happens in the Iris 3D printer made by MCor in Ireland – coloured inks can be sprayed on to create stunning, full-colour output.

Other variants of sheet lamination include the 'ultrasonic additive manufacturing' (UAM) process created by Fabrisonic. This uses high frequency sound waves to fuse layers of metal tape in order to produce robust metal parts. Fabrisonic's technology has the advantage of being able to fuse different metals into the same part, once again allowing the creation of items – such as metal objects with embedded sensors – that could not be created by traditional manufacturing techniques.

DIRECT DIGITAL MANUFACTURING

As we have seen, there are already a large number of 3D printing technologies that can manufacture items using an ever-broadening range of materials. It is therefore already possible to 3D print complex objects made from multiple materials, even if today this requires different components to be output on different printers based on different technologies. But in the future? Well, as local digital manufacturing takes hold, so there is a high probability that we will develop local 'microfactories' or 'distributed manufacturing facilities' (DMFs) that will combine multiple 3D printing technologies to allow small-scale custom manufacture in almost any location.

The creation of multi-technology microfactories probably lies many years and even decades into the future. It is therefore worth considering what kinds of things are most likely to be 3D printed in the shorter-term using existing, single technologies. Today, most things that are 3D printed are product prototypes. But as early as 2020, the majority of industry insiders expect that this will no longer be the case.

As 3D printing matures into a mainstream manufacturing method, its primary area of application will be the fabrication of production tooling. Most traditional production processes rely on dedicated tools, jigs and fixtures to make and assemble product parts, and some of these are starting to be 3D printed. This may not sound that exciting, but nevertheless offers an enormous potential to cut costs, to improve quality, and to allow a greater variety of products to be created.

Volvo Trucks in Lyon, France have recently reduced the time to manufacture some production tools from 26 days to just 2 days by 3D printing them. Meanwhile, case studies published by ExOne – who make 3D printers that manufacture sand molds for use in traditional metal casting – have revealed that mold production lead times can be reduced from months or weeks to days or even hours. The 3D printout of sand cast molds can also reap cost savings as high as 85 per cent.

While the 3D printing of prototypes and tooling can deliver significant cost and other advantages, sometime in the 2020s a tipping-point will be reached and the majority of 3D printed items will be final products. Using 3D printers to make such finished consumer or industrial goods is known as 'direct digital manufacturing' (DDM), which the Society of Manufacturing Engineers define as 'the process of going directly from an electronic digital representation of a part to the final product'.

Although DDM is still in its infancy, in certain circumstances it already offers a wide range of advantages over traditional production methods. The first of these is the ability to reduce the cost of low-run production, and indeed to make possible the manufacture of components that would be prohibitively expensive to create using traditional machining or casting techniques. Stratasys, for example, have reported that 3D printing is already cheaper than injection molding for production runs of less than 5,000 plastic parts. This is because the cost of tooling a mold is saved.

MATERIAL SAVINGS

Another major benefit of 3D printing is the manufacture of products using fewer raw materials. In part this can be achieved because 3D printing is an additive process. This means that 3D printing starts with nothing and adds only the material that is required. This makes 3D printing more resource efficient than subtractive manufacturing processes – such as machining – which start with a lump of material and shave parts of it away.

In addition to reducing the amount of material that ends up on the factory floor, 3D printers can fabricate parts with highly material-efficient geometries that would be impossible to fashion using traditional methods. When molten plastic or metal is injected or poured into a mold, the resultant component inevitably comes out solid. But when things are 3D printed, the insides of components can be made hollow or semi-solid by controlling the level of 'infill' required. All of this means that 3D printing will allow people in the future to make products that are similar to those we have today, but which consume fewer natural resources.

Particularly keen to produce parts that use less material – and which are hence lighter – are the aerospace sector. It is therefore hardly surprising that aerospace companies are at the forefront of DDM. Indeed, the Airbus A350 XWB passenger aircraft already has some brackets that are 3D printed in titanium using powder bed fusion. This has allowed the bracket design to be optimized to a shape that meets structural requirements, but which uses 30 per cent less metal than a traditional milled or cast part. The quantity of metal wasted in the manufacturing process is also reduced from over 95 per cent to around 5 per cent. This improvement in the material 'buy-to-fly' ratio, coupled with the fact that tooling is no longer required, results in cost and time savings of up to 75 per cent.

Also saving time, cost and materials are Winsun Technologies in Shanghai, China. Here buildings are being fabricated using a material extrusion 3D printer that is 6.6 metres high, 10 metres wide and 150 metres long. Winsun's hardware is fed a special 'ink' comprised of cement, glass fiber and construction waste, and accrues material savings partly due to its additive process, and partly because 3D printed walls do not have to be entirely solid to meet structural requirements. In fact, by including air gaps within walls, their insulation properties can actually be improved. Winsun estimates that by 3D printing buildings, it can achieve material savings of between 30 and 60 per cent. The company also estimates that production times can be cut by as much as 70 per cent, with labour costs reduced by up to 80 per cent.

Unlike many other 3D printing pioneers, Winsun has already made good on its amazing claims. For a start, in March 2014, the company 3D printed ten houses in one day. These rudimentary but functional dwellings had a floor area of 200 square metres, and cost about $5,000. Were this not impressive enough, in January 2015 Winsun showcased a five storey 3D printed apartment block, together with a 1,100 square metre 3D printed mansion.

Also in January 2015, Winsun revealed that it had orders for 10 3D printed mansions, as well as 20,000 houses to be 3D printed for the Egyptian government. Winsun additionally announced plans to establish factories in at least 20 countries, including the United States, Saudi Arabia, the UAE, Qatar, Morocco and Tunisia. The company's aim is to provide cheap and efficient homes for low-income families, and I would place a fair bet that it is going to succeed. If, in 20 years time, you are living in a newly constructed building, there has to be a reasonable possibility that it will have been 3D printed.

MASS CUSTOMIZATION

While most factories need to keep costs down by specializing in mass production, a 3D printer need never make the same item more than once. As a consequence, 3D printing is likely to lead to a new age of mass customization, with more and more products personalized to individual requirement.

One of the first pioneers of 3D printed mass customized products was Nervous System, based in Somerville, Massachusetts. Via the company's website at n-e-r-v-o-u-s.com, visitors can 'generate their own 3D printed jewelry' by moving sliders, dragging morph targets, and otherwise sculpting the geometry of a rotatable 3D model. When the personalized design is complete, the user can choose to have it 3D printed in plastic in a range of colours, or cast in a choice of metals from a 3D printed wax pattern.

In addition to web-crafted jewelry, custom 3D printed figurines are now starting to become popular. Several companies – including Twinkind, iMakr, Shapify and the UK supermarket Asda – now offer a service where they scan a person and make a miniature copy using a colour binder jetting 3D printer. Alternatively, over at MyMakie.com, visitors can design their own 3D printed doll, with interactive, onscreen control of everything from clothing and hairstyle, down to mouth shape, ear shape and even nostril flare. The resultant 'Makie' is 3D printed in nylon using powder bed fusion.

3D printing is now also starting to be used in the manufacture of personalized medical appliances. Today, the best things to 3D print are small, customized and expensive, which makes medicine a prime sector for 3D printing application. It is therefore hardly surprising that a 2014 Morgan Stanley Blue Paper found that nearly 40 per cent of new 3D printing patent applications are already in the medical sector.

The range of end-use medical products that have been 3D printed is growing rapidly, and includes dental appliances, hearing aid shells, hip implants and artificial limbs. For example, in 2012 a replacement human jaw bone was 3D printed in titanium for an elderly lady whose own jaw had been damaged by a bone infection. Meanwhile, a Japanese company called Fasotec have developed a material jetting 3D printing process called 'bio-texture modelling'. This allows models of internal organs to be derived from scan data, as illustrated by the liver model shown in figure 1.5. Such custom models are 3D printed from materials that feel organic to the touch, and allow surgeons to plan complex operations before cutting flesh.

Right now, dentistry is at the forefront of medical 3D printing, and by 2025 it is probable that the majority of dental appliances will be fabricated in a few minutes using some form of photopolymerization. In the second half of the next decade, 3D printed replacement hip and other joints are also likely to become common. By the mid 2030s, it is therefore perfectly possible that a billion or more people will have plastic or metal 3D printed prosthesis inside their bodies.

LOCALIZATION & PRODUCT REPAIR

As I indicated at the start of this chapter, the most significant advantage of 3D printing will be the possibility to manufacture products on a local basis. People sometimes tell me that this is a crazy idea, as nobody will be able to make their own furniture on a desktop 3D printer. While this may well be the case, hardware for the local printout of large items is starting to be developed. For example, BigRep in Germany now sell a material extrusion 3D printer called the BigRep ONE. This has a build volume of one cubic metre, and can therefore

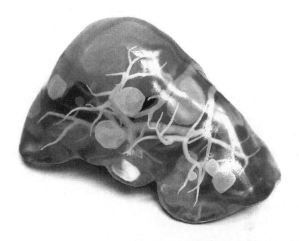

Figure 1.5: 3D Printed Liver Model by Maki Sugimoto/Fasotec.

print a small table or a chair in one piece. The idea of future furniture stores locally 3D printing at least some of their wares is subsequently perfectly plausible. In fact, if a store really wanted to, it could start selling locally 3D printed plastic furniture right now.

Further expanding the build envelope, Cincinnati Incorporated in Ohio have created the Big Area Additive Manufacturing machine (BAAM). This is a material extrusion 3D printer that makes very large items out of a variety of build materials which include ABS plastic reinforced with carbon fiber. The BAAM already comes in two sizes, the largest of which can 3D print objects up to 240 x 93 x 72 inches (6.1 x 2.36 x 1.8 metres) in size.

In September 2014, the BAAM was used by Local Motors to 3D print a one-part chassis, frame and body for an electric vehicle called the Strati. In January 2015, the Oak Ridge National Laboratory (ORNL) next employed the BAAM to 3D print the body of a reproduction Shelby Cobra sports car. The printout of these large sections of both vehicles indicates the potential for the future local production of very complex products. Indeed, as Local Motors explains on its website:

> Gone are the days of mega- or even giga-factories that consume tremendous amounts of time and energy to fabricate products. A more sustainable, nimble and flexible factory is on the horizon. Called microfactories, these diminutive factories drastically change how we produce large consumer goods for unique local needs.

Local Motors plan to open 100 microfactories worldwide by 2025, and already have four in operation. Although these are not currently equipped with BAAM 3D printers (instead vehicles are produced via conventional means), in time it is hoped that 3D printed vehicles like the Strati will be locally produced. While the first Strati took 44 hours to print, Local Motors intends to get this down to 24 hours. It is also already taking orders for the first Stratis to go on general sale.

In addition to allowing products large and small to be manufactured on demand, 3D printing will help to save resources by facilitating increased product repair. For example, if a Strati driver happens to crash their beloved vehicle, they should be able to return it to a microfactory where replacement parts will be replicated and fitted.

Today, many items cannot be repaired because spares are unavailable. But in the future, spares for almost any product should be able to be 3D printed from a downloaded object file or a 3D scan. Airbus is in fact already installing 3D printed spare parts in older aircraft when original replacements are no longer available. Some future products may even come with a portable storage device that contains a complete, digital inventory of replacement parts.

PERSONAL MANUFACTURING

Eagerly awaiting the above kinds of development are a growing number of techno-savvy individuals collectively known as the 'Maker Movement'. A 'maker' is simply the latest term for somebody who wants to make or mend their own stuff, and who in such pursuit becomes skilled in the ways of DIY or 'DIWO' (do-it-with-others). Supporting the latter are a growing number of local community 'Fab Labs' and 'hackerspaces'. These allow people to learn construction and repair skills from each other, and to share specialist hardware like 3D printers.

Today there are nearly 900 Fab Labs and hackerspaces worldwide, and it is not hard to imagine how they will evolve into the first generation of local microfactories. Already personal 3D printers can allow some things to be personally fabricated in people's own homes. However, I suspect that for a few decades at least, most 3D printed products will be produced in microfactories – or in more conventionally-scaled production facilities – that will be run by single manufacturers, single retailers, local manufacturer or retailer conglomerates, or local community groups.

Options for home-based digital manufacturing are likely to remain limited for some time to come. This is because most products will only be able to be 3D printed on hardware that will remain bulky and very expensive until at least the 2030s. 3D printers capable of outputting objects in multiple materials, including metals, are also set to remain too complex for the majority of people to operate and maintain in their own homes until at least the mid-2020s. Granted, by the early 2020s, a large minority of households are likely to own some kind of domestic 3D printer. But these relatively small, cheap fabricators will have limited production capabilities. School projects, plant pots, buttons, shoes, toys and haberdashery will early next decade be liberally springing from domestic 3D printers. However, when you want a new car or washing machine, you will have to go to a microfactory run by Local Motors, Local Domestic Appliances, or a local community group.

BIOPRINTING

So far in this chapter I have focused on 3D printing technologies and applications that turn digital designs into solid but inorganic reality. There is, however, already an entirely distinct variant of 3D printing that builds objects from living cells. This quite wondrous innovation is known as 'bioprinting', and has the potential to transform not just medicine, but all future forms of local digital manufacture.

One of the first bioprinting pioneers was a Japanese paediatrician called Makoto Nakamura. Back in 2002, the good doctor realised that the droplets of ink jetted out by a standard 2D photo printer were about the same size as human cells. He subsequently began experiments to modify a standard Epson photoprinter to allow it to print human tissue. This he eventually achieved by encasing the cells in a sodium alginate hydrogel to stop them from drying out, and by jetting them into a calcium-chloride solution.

Continuing his work, between April 2005 and March 2008 Nakamura led a project at the Kanagawa Science and Technology Academy that scratch-built an experimental bioprinter. This was capable of creating 'biotubing' from layers of two different types of cells. In the future, such biotubing could be used by surgeons as a living replacement for human blood vessels.

Another bioprinting pioneer is Gabor Forgacs from the University of Missouri. In 1996, Forgacs recognised that cells stick together during embryonic development in a manner that could greatly assist in artificial tissue fabrication. By 2004, Forgacs had used this insight to develop a bioprinting technology that fabricates tissues not from individually jetted cells, but by building up layers of 'bio-ink

3D PRINTING

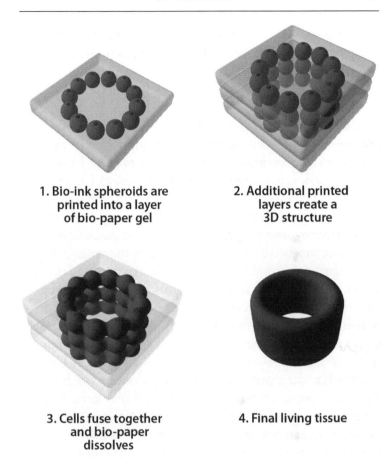

Figure 1.6: Bioprinting with Bio-ink & Bio-paper.

spheroids' that each contain tens of thousands of cells. The cells in question are cultured in the lab from samples obtained via human biopsies.

In 2007, Forgacs founded a company called Organovo, and by 2008 had developed a prototype bioprinter. As shown in figure 1.6, this injects 'bio-ink spheroids' into layer-upon-layer of a water-based 'bio-paper' support structure. Nature then takes control of the fabrication process, and over many tens of hours the cells contained in the bio-ink spheroids slowly fuse together. With the bio-paper support dissolved or otherwise removed, what eventually results is integrated, living tissue.

The really amazing aspect of this kind of bioprinting is that, during the 'maturation phase', the cells contained in the bio-ink spheroids not only fuse into solid tissue, but also rearrange. For example, a blood vessel may be bioprinted from bio-ink spheroids that contain an aggregate of primary endothelial cells, smooth muscle cells and fibroblasts. After printout, these different cell types will have been randomly positioned by the bioprinter. But during maturation, the primary endothelial cells migrate to form the inner lining of the bioprinted blood vessel, while the smooth muscle cells travel to the middle, and the fibroblasts move into place to constitute the blood vessel's outer tissue.

Because it outputs bio-ink tissue spheroids, Organovo's technology can be much faster than a bioprinting technique that jets individual cells. Moreover, as the cells reposition themselves after printout, it is not necessary to bioprint every last detail of the organ or tissue under construction. Rather, natural organic processes can be left to form complex, intricate structures such as the capillary networks within bioprinted kidneys. Print something in a traditional plastic or metal, and what you get is exactly what you print. But when an object is bioprinted, the fabrication process can continue after the print head has finished its activities.

In 2009, Organovo developed the Novogen MMX as the world's first commercial bioprinter. In December 2010, this was subsequently used to create the first bioprinted human blood vessels, and since that time has also managed to fabricate small samples of human skeletal muscle, bone and liver tissue. In fact, in November 2014, Organovo actually started selling a bioprinted human liver product called exVive3D. This was the world's first commercial bioprinted tissue, and is fabricated for use in drug testing. As Organovo further explain:

> Organovo's exVive3D Liver Models are bioprinted, living 3D human liver tissues consisting of primary human hepatocytes, stellate, and endothelial cell types, which are found in native human liver. The exVive3D Liver Models are created using Organovo's proprietary 3D bioprinting technology that builds functional living tissues containing precise and reproducible architecture. The tissues are functional and stable for at least 42 days, which enables assessment of drug effects over study durations . . . well beyond those offered by industry-standard 2D liver cell culture systems.

While no company or research team has yet managed to 3D print tissues or organs for human transplantation, it is only a matter of time before this occurs. The timescale is hard to predict, but bioprinted arterial and nerve grafts, heart valves, and substitutes for 'simple' organs like kidneys, are probably a realistic possibility in the late 2020s or early 2030s. Beyond that, and as I conceptualize in figure 1.7, by the 2040s we could see bioprinters capable of fabricating complex organs like the human heart.

Bioprinting large organs will not be easy, and not least due to the problems inherent in keeping sizable bioprintouts alive during fabrication. Just one researcher working on this 'thick tissue problem' is Ibrahim Ozbolat at the Biomanufacturing Laboratory in the University of Iowa. As he explained to me back in 2013:

> Systems must be developed to transport nutrients, growth factors and oxygen to cells while extracting waste so the cells can grow and fuse together,

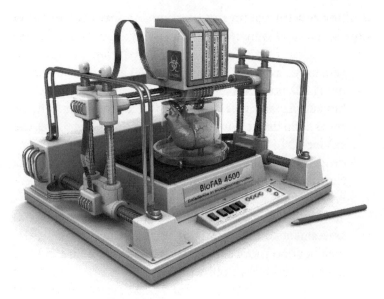

Figure 1.7: Bioprinter Concept. Image: Christopher Barnatt.

forming the organ. Cells in a large 3D organ structure cannot maintain their metabolic functions without this ability, which is traditionally provided by blood vessels.

Further organizations working to make the 3D printout of human tissue an everyday occurrence include Regenovo in China, Cyfuse Biomedical K.K. in Japan, and the Wake Forest Institute in Winston-Salem, NC. The latter operates under the directorship of Anthony Atala, who in 2006 grew several bladders on scaffolds in his lab, before successfully transplanting them into patients who are still alive.

Since 2008, the Wake Forest Institute has led or co-led an interdisciplinary network called the Armed Forces Institute of Regenerative Medicine (AFIRM), and in this capacity is conducting a range of bioprinting initiatives. These include working toward the future printout of bone, nerves, blood vessels, fat and muscle for use in the facial and skull reconstruction of wounded soldiers. Experiments are also progressing to allow burns and other wounds to be healed directly with a form of in situ bioprinting. Such future 3D printing technology would scan a wound, remove damaged cells, and then print layers of new cells directly onto the human body.

Already a laser scanner has been used to produce 3D models of test injuries inflicted on mice. Data from these models was then used to guide print heads that sprayed fibroblast cells (which make up the deep layers of the skin) and keratinocyte cells (which compose the top skin layer) onto the rodents' wounds. After printout, maturation took place, with the cells fusing into new skin in two or three weeks, compared to the five or six weeks that it took wounds in a control group to heal naturally.

Wake Forest hope to be using their in situ technology on wounded soldiers by 2020. And if this goal is met, the foundation may well have been laid for a medical revolution. After all, when and if it becomes possible to bioprint directly onto the body, keyhole surgical tools could be developed that will be able to be inserted into a patient to replace damaged organs (or parts thereof) from the inside. Used in combination with other cutting edge medical technologies, in situ bioprinting may one day even permit a radical re-engineering of the human body. We may, for example, learn to fabricate novel organs, such as gills to allow people to breath underwater.

Another extraordinary possibility may be the creation of a face printer. As pictured in figure 1.8, such future hardware may be able to remove existing flesh and bone, before replacing them with layer-after-layer of fresh cells. Such a development lies decades into the future, and may not be to everybody's taste. This said, in May 2015 bioprinting pioneer

Figure 1.8: Face Printer Concept. Image: Christopher Barnatt.

Organovo announced a research partnership with L'Oreal to develop bioprinted skin tissue. As Guive Balooch, the Global Vice President of L'Oreal's Technology Incubator, enthused:

> We developed our technology incubator to uncover disruptive innovations across industries that have the potential to transform the beauty business. Organovo has broken new ground with 3D bioprinting, an area that complements L'Oreal's pioneering work in the research and application of reconstructed skin for the past 30 years. Our partnership will not only bring about new advanced in vitro methods for evaluating product safety and performance, but the potential for where this new field of technology and research can take us is boundless.

LDM HORIZONS

As we have explored in this chapter, 3D printing is a technology with a very wide range of future applications, and which is developing at a startling pace. 3D printed prototypes, aerospace parts, medical devices, buildings, toys and living tissue are already a reality, with the majority of the population destined to own at least some 3D printed things by the turn of the decade. It is therefore not surprising that in March 2015, the Chinese Government announced its *Additive Manufacturing Industry Promotion Plan* to boost national competence and confidence in 3D printing. Make no mistake, even by 2020, 3D printing will have radically morphed from being a niche rapid prototyping technology into a mainstream industrial process of strategic importance to many nations.

The above point noted, it is critical to appreciate that by the time that 3D printing really enters the mainstream, it will have converged so greatly with other technologies that the label '3D printing' is unlikely to be in general usage. Not least, as we have seen in this chapter, new material innovations are already a critical aspect of 3D printing development, and the opportunity to use a biological maturation process to help create non-living printouts out of new materials presents an enormous future opportunity. Today, bioprinting is being developed specifically to produce organic printouts that will continue to live as viable human tissue. But there is the related possibility of creating bioprinters that will output synthetic living materials that will mature into their final, programmed form after printout and then die on digital cue.

For millennia human beings have fashioned products from natural organic materials – such as leather, wood and bone – that were once alive, but which we use as manufacturing inputs in a dead state after we have harvested them. Future bioprinting innovations could allow digital fabrication using an entirely new range of synthetic, organic materials that remain alive throughout a production (3D printing) and post-production (maturation) process. Imagine, for example, a 3D (bio)printer that could output layers of an organic material akin to living coral, with the end result of its fabrication process being a complex and (largely) dead form like a coral reef. 3D printing in concrete may help us create more material-efficient buildings. But it will never compete in environmental terms with a future bioprinting process capable of turning a digital design into a building that is actually part of the 'natural' world.

To open our eyes to the far broader avenues for digital fabrication that lie ahead, the next two chapters detail potential developments in synthetic biology and nanotechnology. The former involves the artificial creation or radical redesign of life itself, while the latter takes us 'beyond layers' to consider future manufacturing on a molecular scale. To build the local 'microfactories' or 'distributed manufacturing facilities' predicted in this chapter, we will almost inevitably have to merge 3D printing with synthetic biology and nanotechnology, and probably with a good slug of artificial intelligence and android robotics also cast into the mix. Local digital manufacturing is I am certain going to be a survival-critical future development for our species. But it will take far more than 3D printing to turn it into the Next Big Thing.

2
SYNTHETIC BIOLOGY

Most of the manufacturing innovations of the 20th century utilized inorganic technologies. In contrast, an increasing proportion of new 21st century fabrication methods are likely to depend on organic processes. The most sophisticated computer ever mass produced is, in fact, already organically fabricated. This amazing computational hardware is generally known as the human brain, and provides a useful reminder of what biological productive systems are capable of.

As we saw in the last chapter, scientists have already created first generation bioprinters that can 3D print layers of living cells. As this cutting-edge technology is developed, within a few decades we will be able to bioprint arterial and nerve grafts, kidneys and even replacement human hearts. Bioprinting subsequently has the potential to reduce organ transplant waiting lists to zero, and could even permit the redesign of the human form to cosmetic whim. In parallel, derivatives of bioprinting will allow future products to be crafted from revolutionary new organic materials. And yet, a few centuries from now, bioprinting will probably be remembered as a rather quaint and second-best biological fabrication method.

Any bioprinting process relies on a supply of cells that are initially grown in a lab. A bioprinter is then loaded with these cells, which are subsequently output in layers according to the dictates of a 3D computer model. During a post-printout maturation phase, the cells then fuse and rearrange to form viable living tissue.

While the aforementioned process is quite amazing, it could also be considered needlessly cumbersome. Every cell being bioprinted does, after all, contain a digital code that could potentially tell it how to form into the tissue structure being fabricated. So rather than loading a stock of cultured cells into a bioprinter, and then 3D printing them in layers, it would be better to create the correct conditions to allow the cells to grow into desired tissue structures entirely in accordance with their internal biological programming.

The best technology for fabricating anything from living cells has to be life itself. A baby does not grow its liver and other internal organs by supplying a cell sample to a bioprinter. Rather, it lets its genetic code and self-assembly processes get on with replicating and positioning the necessary cells in just the right place. If we could control this kind of extraordinary activity, we could artificially create organs for transplant – not to mention food, medicines, fuels, clothing and all kinds of other useful things – using 'natural' biological processes. This may sound absolutely crazy. But it is what an amazing new science called synthetic biology is all about.

Synthetic biology uses engineering principles to design and construct living things. By allowing the creation of 'synthetically modified organisms' (SMOs), synthetic biology will provide us with revolutionary fabrication methods and other biological technologies that would never have evolved in the natural world. Or as German pioneer Intrexon more formally explains:

> Synthetic Biology is the engineering of biological systems to enable rational, design-based control of cellular function for a specific purpose. The programming of DNA and reformatting of genetic cir-

cuitry within cell platforms has created a paradigm shift whereby the analysis of biology is being supplanted by its synthesis. This advancement has the potential to significantly impact approaches relied upon for some time across a variety of industries.

Today, most of the things that we buy are inorganically manufactured rather than biologically grown. Yet in a few decades time, local biofactories may routinely cultivate fuels, plastics, medicines, bioelectronics and potentially all manner of industrial and domestic goods. In fact, according to a 2014 report from BCC Research, the global market for synthetic biology will be worth nearly $11.9 billion annually as early as 2018. This already makes the predicted synthetic biology marketplace larger than the more widely promoted arena of 3D printing, with the latter unlikely to hit a global turnover of $10 billion until around 2019 or 2020.

Interest and investment in synthetic biology are also growing rapidly. For example, according to a September 2015 report from the Synthetic Biology Project at the Wilson Center in Washington, DC, the United States government invested over $820 million in synthetic biology research programs between 2008 and 2014. In fact, in 2014, the Defence Advanced Research Projects Agency (DARPA) alone spent more than $100 million on synthetic biology projects. There is therefore good reason to believe that synthetic biology will be a very important Next Big Thing.

GENETIC FOUNDATIONS

For many years scientists have been investigating the biological building blocks that constitute life. Back in 1869, Swiss chemist Johann Friedrich Miescher revealed that the chemical code of living things was stored in a substance that we now term 'deoxyribonucleic acid' or more commonly 'DNA'. In 1944, a team led by Oswald Avery at Rockefeller

University then demonstrated that genes consist of distinct portions of DNA, while in 1953 the double-helix structure of DNA was discovered by James Watson and Francis Crick. Following these ground-breaking achievements, it was almost inevitable that other researchers would attempt not just to read the code of life, but to re-write it to human specification.

Rising to the challenge, in 1973 Stanley Cohen and Herbert Boyer created the first 'transgenic' organism when they chemically spliced a gene from one bacterium into the DNA of another. Three years later, Boyer co-founded a company called Genentech, which only two years after that had managed to synthesize human insulin by inserting the appropriate gene into an *E. coli* bacterium.

Other pioneers of genetic modification (GM) included a company called Calgene, which in 1994 launched its Flavr Savr tomato. This was genetically modified using 'gene-silencing' technology which shut down the gene that causes tomatoes to release their seeds and rot. As a consequence, the Flavr Savr can be ripened for longer without the risk of the fruit being crushed on the way to market.

Following in Calgene's footsteps, from 1995 Monsanto introduced a range of corn and other GM crops. These were initially modified to be resistant to particular pesticides, so enabling farmers to kill weeds more easily without damaging their harvests. Today, GM animals have also been commercially created, most notably including the AquaAdvantage salmon engineered by AquaBounty Technologies in the United States. This has had ocean pout and chinook salmon genes spliced into its DNA, which causes the fish to grow to full size in 18 rather than 36 months.

The genetic modification of plants and animals remains a controversial topic, with products like the AquaAdvantage salmon not yet licensed for human consumption. Even so, at the time of writing, around 90 per cent of the corn, cotton and soybeans planted in the United States are genetically modified. To a more limited extent, GM crops are also being grown in 26 other countries, most notably including Argentina, Brazil, India and China. Human 'interference' in 'natural' biological processes is therefore a well established practice. This said, to date it has been a practice that has only involved the minor alteration of existing species.

BIOLOGICAL ENGINEERING

The cultivation of GM crops and even GM animals may in the future help to improve food security. Nevertheless, to harness the full potential of life as a manufacturing technology, we will need to progress far beyond making minor genetic tweaks to existing animals, plants and micro-organisms. This brings us into the new realm of synthetic biology, which standardizes and assembles the components of life in radically new ways.

The first researchers to envisage the creation of biological systems from molecular components were Francois Jacob and Jacques Monod in 1961. Unfortunately, at the time the genetic knowledge and tools to achieve this did not exist. It was therefore not until the mid-1990s, when the aforementioned developments in genetic modification began to take place, that an engineering mentality actually started to be applied to biology. Slowly confidence and competence grew, and in 2004 the first international conference on synthetic biology – SB1.0 – was held at the Massachusetts Institute of Technology (MIT) in the United States.

SB1.0 brought together researchers from molecular biology, chemistry, engineering, physics and computer science, and established synthetic biology as a stand-alone discipline. In particular, the conference firmly established the idea of designing or redesigning biological parts, devices and systems using engineering principles. This bedrock concept distinctly separates synthetic biology from 'traditional' genetic engineering, as the latter 'merely' seeks to alter one or a few characteristics of an existing form of life.

Given that synthetic biology is a child of the Internet Age, its development has consistently been catalyzed by online communities who share open web resources. For example, over at SyntheticBiology.org you can find a 'group of individuals, groups and labs from various institutions who are committed to engineering biology in an open and ethical manner'. To this end, SyntheticBiology.org are working toward the 'design and construction of new biological parts, devices and systems', together with 'the re-design of existing, natural biological systems for useful purposes'.

GENETIC LEGO

All forms of engineering break things down into standardized components, so enabling complex systems to be understood, assembled and repaired in a modular fashion. Before synthetic biology introduced such an approach to the biosciences, the potential scratch design of biological mechanisms lay well beyond human skill and comprehension. But with biology reduced to a set of standardized 'Lego blocks', so the conceptual and practical construction of novel biological systems started to become a possibility.

The significance of introducing the engineering principle of standardization into the biosciences cannot be overstated. The magnitude of the innovation is also captured to great effect in a 2014 report from the European Research Area Network in Synthetic Biology (ERASynBio). As Professor Victor de Lorenzo explains in the *Foreword* to the report, synthetic biology:

> . . . brings about a fresh way of looking at living systems, not as complex objects to be thoroughly understood, but as sources of amazing building blocks that can be retrieved from their natural

context, reshaped, standardized to fit a given specification and used for a purpose different from their original raison d'être. The transformative potential of this simple principle is extraordinary, perhaps only comparable to the development of the steam engine in the 18th century. If this invention marked the start of domesticating physical energy for humanity's benefit that began our modern era, synthetic biology will enable a new type of industry – and ultimately a society where biological agents and materials (from fuels and environmental catalysts to intelligent fabrics and smart therapeutic agents) will take over many of the roles that are currently assigned to far more primitive and inefficient counterparts.

The first major attempt to create a modular collection of biological building blocks was the Registry of Standard Biological Parts (RSBP). This was developed to digitally catalogue and physically store DNA components called 'BioBricks'. The RSBP was initially based at MIT, though in January 2012 it was spun off to be run by the International Genetically Engineered Machine Foundation or 'iGEM'. This independent, non-profit organization is dedicated to the advancement of synthetic biology, and now hosts the RSBP at parts.igem.org.

iGEM describes its registry as 'a growing collection of genetic parts that can be mixed and matched to build synthetic biology devices and systems'. Every year iGEM holds a synthetic biology competition that encourages the development and sharing of new BioBricks. This is also very popular, with over 1,500 new genetic components submitted to the RSBP BioBrick catalogue in 2014.

While the creation and continued operation of the RSBP remains important, in recent years new methods for the assembly of genetic components have broadened the focus of synthetic biology toward the establishment of standardized languages. For example, the 'synthetic biology open language' (SBOL) now provides a standard data format for describing biological parts, so facilitating their modular exchange. Open software for the design of biological parts and systems is also being created, including GenoCAD. As the website for the latter explains:

> GenoCAD is an open-source computer-assisted-design (CAD) application for synthetic biology. The foundation of GenoCAD is to consider DNA as a language to program synthetic biological systems. GenoCAD includes a large database of annotated genetic parts which are the words of the language. GenoCAD also includes design rules describing how parts should be combined in genetic constructs. These rules are used to build a wizard that guides users through the process of designing complex genetic constructs and artificial gene networks.

The above sounds most impressive and *actually is* very impressive indeed. This said, you may by now be wondering what standardized, modular biology may practically achieve. Well, for a start, synthetic biologists have already created a range of 'gene circuits', including biological toggle switches. In 2013, a team of bioengineers from Stanford University even synthesized a biological transistor that they named a 'transcriptor'. In nature, such biological equivalents of electronic components perform genetic logic to regulate activities like cell replication. But, in the future, synthetic gene circuits could facilitate the development of biological computers based on 'wetware' akin to the human brain.

Future advancements in gene circuit design could mean that, 30 years from now, the virtual assistant embodied in your smartphone or robot may actually have a biological central processing unit and living memory. Such future computing devices could therefore grow hardware upgrades as required, and may be sustained with chemical feedstocks rather than an electrical supply.

In addition to gene circuits, the pioneers of synthetic biology have also managed to fashion several novel microorganisms, as well as a few synthetic higher forms of life. Over the following pages we will explore these developments, starting with synthetic bacteria, moving on to future re-engineered plants, and finally the first biofactories to roam around on four legs.

THE BIRTH OF SYNTHETIC LIFE

While the mainstream, conceptual foundations of synthetic biology were laid at the SB1.0 conference in June 2004, the new discipline had to wait six more years to really come of age. This great day occurred on 20 May 2010, when the J. Craig Venter Institute (JCVI) published work that described how it had successfully constructed the first self-replicating, synthetic bacterial cell. This they labelled *Mycoplasma mycoides* JCVI-syn1.0, although the press christened it 'Synthia'.

The genome for JCVI-syn1.0 was constructed from 1,078 'cassettes of DNA' that the JCVI team designed and spliced together into a 1.08 million DNA base pair sequence. This entirely synthetic genome was then transplanted into an existing *Mycoplasma capricolum* bacterium and electrically booted-up in the JCVI lab to create a new form of life.

As JCVI proclaimed, the creation of JCVI-syn1.0 proved that 'genomes can be designed in the computer, chemically made in the laboratory and transplanted into a recipient cell to produce a new self-replicating cell controlled only by the synthetic genome'. Or as the company additionally enthused, while for years scientists had known how to turn the chemical sequences held in DNA into computer code, only with the creation of JCVI-syn1.0 had anybody actually managed to reverse the process and use binary data to program the characteristics of a living cell. It is therefore not surprising that J. Craig Venter described JCVI-syn1.0 as 'the first self-replicating life form with a computer as its parent'.

JCVI-syn1.0 took 15 years to fabricate, and built on the JCVI's success in creating the world's first synthetic virus in 2003, and the first synthetic bacterial genome in 2008 (which at the time they could not boot-up or 'activate' to become a living cell). During the creation of JCVI-syn1.0, JCVI even developed its own alphabetic code based on the four DNA chemical bases A, C, G and T. This was then used to write watermarks into their synthetic genome's DNA, one of which was an e-mail address.

While JCVI is a non-profit organization, its funding to create the first synthetic bacterium came from a fully commercial sister company called Synthetic Genomics Inc (SGI). Building on JVCI's work, SGI is subsequently planning:

> . . . to create the next generation of renewable and sustainably-produced biology-based products. From new vaccines and therapeutics, food and nutritional products, humanized organs for transplant, biofuels, biobased-chemicals, and agricultural solutions, we are producing products through our own programs and with industry leading partners. We believe society can depend on science to alleviate many of our current issues, and SGI is blazing the trail to turn innovative science into life-changing solutions. Our imagination is our only limitation and we imagine a world where synthetic biology will transform the world.

FERMENTING BIOPLASTICS

Similarly working to use synthetic biology in the production of biochemicals are Yu Kyung Jung & Sang Yup Lee from the Korean Advanced Institute of Science & Technology (KAIST). Specifically, these two researchers are working on a novel method for creating the bioplastic polylactic acid (PLA), and have already reported very promising results.

In the face of inevitable future oil shortages, PLA is a substitute for petroleum-based plastics that we may rely on significantly in the decades ahead. At present, PLA is almost entirely produced from a starch-rich agricultural feedstock, such as corn. Unfortunately, however, the industrial processes involved are currently complex and not that environmentally friendly. This is because they not only involve the fermentation of lactic acid from a feedstock, but in addition a chemical post processing stage to transform the acid into the long polymer chains from which plastics are made.

Aware of the limitations of existing PLA production methods, Jung and Lee have engineered a synthetic *E. coli* bacterium that can directly ferment PLA from glucose. As you may remember from the last chapter, PLA is already fairly widely used as a 3D printing filament. Jung and Lee's groundbreaking work hence paves the way for the production of a key future raw material in any local biofactory with access to sugar beet, sugar cane, or any other glucose source. The latter potentially include algaes, which have a high sugar and oil content, and can be very fast growing. Ten years from now, it is therefore reasonable to predict that algae could be grown locally in vats, and then fermented into PLA or other bioplastics as a 3D printing raw material. Progressing us in this direction, a company called Algix already sell a PLA filament that is 20 per cent algae based.

FROM 100,000 YEARS TO A FEW DAYS

Another notable synthetic biology pioneer is Amyris, who are proving very successful in engineering organisms 'to address some of our planet's most daunting problems'. As the company explains, its scientists:

> ... have developed genetic engineering and screening technologies that enable [them] to modify the way micro-organisms process sugar. By controlling their metabolic pathways, [they] design microbes, primarily yeast, and use them as living factories in fermentation processes to convert plant-sourced sugars into target molecules.

One of the biochemicals that Amyris produces using its synthetically engineered yeast is farnesene. This is the same oil that provides a waterproof coating on the outside of apples, and is excreted by Amyris' synthetic yeast cells as they ferment sugar cane. The farnesene is then used in a range of products, including a biodiesel called BioFene.

While it naturally takes at least 100,000 years to form oil from fossilised organisms, Amyris is now fermenting farnesene from organic matter in just a few days. The company's first purpose-designed manufacturing facility was opened in December 2012 in Brotas, Sao Paulo in Brazil, and produces farnesene on an industrial scale.

Biofene is a pure hydrocarbon with properties 'superior to those of petroleum-sourced diesel, allowing it to be used as a drop-in replacement in practically any diesel engine'. Known in Brazil as Diesel de Cana, it emits 80 per cent less greenhouse gas emissions than fossil fuel alternatives, and also improves air quality as it results in fewer emissions of particulate matter. At the time of writing, Biofene is being used to run 400 public buses in Sao Paulo, with the fleet having logged over 40 million kilometres while burning it.

In partnership with Total, Amyris have also applied their synthetic biological know-how to manufacture an aviation fuel. This was first used to power a commercial jet aircraft in 2012, obtained regulatory approval in 2014, and is currently being sold to airlines around the globe.

Additional Amyris products already on the market, and similarly fermented from sugar cane, include Neossance Squalane. This is naturally present in the skin, and is sold as a moisturizer in the cosmetics industry. Amyris is also using synthetic biology to help produce lubricants and new ingredients for flavours and fragrances. The company is in addition working with Michelin and Braskem to ferment isoprene from plant sugars for use in the production of tyres. This last endeavour looks extremely promising, and it is hence not surprising to learn that Goodyear is working with DuPont on a similar initiative.

Other commercial synthetic biology pioneers include Life Technologies, Intrexon, Modular Genetics and Solazyme. While Life Technologies, Modular Genetics and Intrexon have very wide-ranging ambitions, Solazyme is focused on re-engineering microalgaes – which it terms 'the world's original oil producers' – so that they can turn sugar feedstocks into 'sustainable, high-performance products'. Just one customer is Belgium soap-maker Ecover, who in April 2014 revealed that it had started to use a synthetic Solazyme algae oil in its products.

While most current synthetic biology pioneers are intent on producing alternatives to conventional chemicals and fuels, a few are focused on the synthetic replication of natural materials that cannot be harvested in quantity. One of the best examples of such a company is Spiber in Sweden, who have engineered synthetic cells that can manufacture spider silk. This natural fiber is well known for its extraordinary mechanical properties, with 'dragline silk' being one of the toughest materials known to exist.

To produce spider silk synthetically, Spiber have isolated a portion of the gene that spiders use to make it, which they term the 'mini-spidroin'. This they have introduced into an *E. coli* bacteria that can be cultivated in bioreactors. In time, Spiber's synthetic spider silk may have a range of applications. For example, it could be used as a surgical twine that will be readily accepted by the human body.

Another rare, natural product with very valuable properties is artemisinin, which is obtained from a plant called *Artemisia annua*, also known as sweet wormwood. Artemisinin is a traditional Chinese herbal medicine that releases oxygen-based free radicals that are very effective in killing the malaria parasite. Given that this parasite has developed a resistance to other medicines, this makes artemisinin a critical antimalarial drug.

Unfortunately, cultivating and harvesting artemisinin from *Artemisia annua* is very labour intensive and expensive as the plant grows slowly and yields are low. In 2004, the Institute for OneWorld Health (IOWH), the University of California and Amyris therefore began working on the Artemisinin Project. This set out to fabricate a semi-synthetic version of artemisinin using synthetic biology, and obtained substantial grants from the Bill & Melinda Gates Foundation. The project also proved a success, with semi-synthetic artemisinin now in large-scale production. The new drug is made from artemisinic acid that is produced by one of Amyris' re-engineered yeast strains, and in 2014 was used in around 150 million antimalarial treatments. In time, the hope is to deliver 300 million treatments annually, and to prevent over 600,000 people dying from malaria every year.

A GROWING INDUSTRY

If the previous few pages have taught us anything, it is that the biotech industry is going to expand very rapidly. For centuries we have been fermenting products including beer, cheese and yoghurt, with the total output of the biotech sector already worth around $250 billion a year. Given the potential to use synthetically engineered micro-organisms to fabricate an extremely wide range of future bioproducts, it is easy to predict a trillion dollar industrial biotech sector sometime in the late 2020s. Kits for the home-brewing of paints and plastics under your stairs or in your garage also have to be a realistic possibility, with 'DIYbio' already being fairly widely discussed online.

By the late 2020s, we will be using synthetic microbes to ferment a wide range of biofuels and other biochemicals. This important point noted, there will be far more to the Synthetic Biology Revolution than the use of re-engineered micro-organisms. We may have started by making designer yeast and bacteria, but this is only because micro-organisms are the easiest forms of life to artificially program. As we become more skilled in assembling standardized biological building blocks, I therefore strongly suspect that we will begin to routinely fabricate both synthetic plants and synthetic animals.

Fairly soon we are all likely to be drinking from bottles manufactured from synthetic bioplastics. But why should we stop at fermenting bioplastics that we then craft into bottles in traditional, energy-guzzling factories? As synthetic biology develops, in the long-term it would make far more sense to engineer synthetic plants capable of growing finished bottles on the vine. And why even cease innovating at empty bottles? If future synthetic plants can be created capable of growing bioplastic containers, then why not go the whole hog and program them to fabricate bottles already filled with a soft drink? In a similar vein, given that we are already garnering expertise in producing isoprene using synthetic microbes, in time there may also be the possibility to engineer amazing new plants from which we could directly harvest car tyres, knicker elastic and rubber bands.

The above kinds of innovations lie decades into the future. Even so, the creation of synthetic plants for productive purposes is already starting to be taken very seriously indeed. Not least, the OpenPlant Initiative has been established by the University of Cambridge in collaboration with the John Innes Centre. This is funded by the United Kingdom's Biotechnology and Biological Sciences Research Council (BBSRC) and the Engineering and Physical Sciences Research Council (EPSRC), and plans to use 'plant synthetic biology' to help create a more sustainable world. Or as the project's homepage at openplant.org further explains:

> Synthetic biology offers the prospect of reprogrammed biological systems for improved and sustainable bioproduction. While early efforts in the field have been directed at microbes, the engineering of plant systems provides even greater potential benefits. In contrast to microbes, plants are already globally cultivated at extremely low cost, harvested on the giga-tonne scale, and routinely used to produce the widest range of biostuffs, from fibres, wood, oils, sugar, fine chemicals [and] drugs to food. Plants are genetically facile, and GM plants are currently grown on the >100 million hectare scale. Plant systems are ripe for synthetic biology, and any improvement in the ability to reprogram metabolic pathways or plant architecture will have far-reaching consequences.

Given that in many countries the cultivation of GM crops has met with a great deal of public resistance, those seeking to create synthetically modified (SM) plants will need to tread very carefully on their path from research lab to commercial harvest. Potentially for this reason, the development of SM plants may occur in two distinct phases. The first of these could involve the development of new plant species created specifically to be fed to synthetic microbes that would in turn ferment them into bioproducts. The second, more radical phase would then focus on engineering SM plants capable of directly producing biochemicals and other bioproducts. Today, we are quite used to farmers harvesting fruit, vegetables, wood, cotton and even chemicals like latex. But by 2030, it is possible that the majority of chemicals could be on tap from the vine. We may even be able to visit a local farm to pick not just our own strawberries, but also organic tupperware, organic shoes, and one day even bio-electronic products like batteries or smartphones.

FACTORIES ON LEGS

While the engineering of SM plants lies years into the future, some research teams have already created the first 'factories on legs'. Most famously, working with the former Canadian company Nexia Biotechnologies, Professor Randy Lewis has created a new species of transgenic 'spider goat'. Now living on a farm at Utah State University, these animals roam around just like 'natural' goats, save for the fact that they have had a spider gene inserted into their DNA. This causes the goats to secrete a spider silk protein in their milk. Back in the lab, this protein can be filtered from the rest of the milk, drawn out into a continuous strand, and wound onto a reel. The resultant biomaterial is stronger than Kevlar, has an elasticity greater than rubber, and may be stitched or woven in a conventional manner.

Initial applications envisaged for spider-goat silk include the production of hardier sports gear, parachute chords, boat sails, airbags, and replacement human ligaments and tendons. As you may recall from a few pages back, over in Sweden Spiber have also created synthetic spider silk using re-engineered *E. coli* bacteria. But as Randy Lewis explains, it is more productive to produce the spider silk protein in goat's milk than in cultured *E. coli*. Given this assertion, there is the possibility that we will increasingly use SM animals to produce bioproducts, rather than relying on micro-organisms cultured in chemical vats.

Engineering and then naturally breeding new species of animals as 'factories on legs' will raise a whole host of future possibilities and concerns. After all, once animals become biofactories, for the first time industrial 'production technology' will be able to be bred and farmed on a local basis almost anywhere on the planet. Back in the last chapter, the 'microfactories' or 'distributed manufacturing facilities' envisaged by companies such as Local Motors were small machine shops equipped with 3D printers. But they could just as easily – or in tandem – turn out to be farms stocked with SM herds.

The engineering of new species that lactate or urinate biochemicals will inevitably raise questions related to inbreeding and any possible contamination of the food chain. It is potentially not that difficult to keep apart different strains of synthetic micro-organism cultivated in a biofactory. But containing large, farmed herds could be quite a different matter if goats, cows and other animals start to be routinely used as factories on legs.

And just what happens if a goat that lactates spider silk mates with a goat from another SM species whose females lactate a bioplastic or biofuel protein? Will future farmers be able to fabricate new materials simply via selective breeding? And will the interface of SM animals and their natural counterparts be able to be controlled?

In the future, we may even use SM animals to produce more than liquid chemicals. For example, in May 2014, Synthetic Genomics entered into an agreement with Lung Biotechnology to develop humanized pig organs. The idea is that Synthetic Genomics will engineer pig cells with modified genomes that will include human DNA sequences. In turn, this will result in new species of pigs that will be born with 'humanized lungs'. Such lungs, and in time other organs, could then be used for organ transplants (assuming that no local bioprinter is available). Sometime in the 2020s, farms may even raise humanized pigs with personalized DNA, so allowing those with appropriate medical insurance to have a set of replacement parts in live storage just in case they ever suffer a major accident.

One day we may even create animals to grow products other than humanized organs. For example, if future computers or robots have bioelectronic brains, then perhaps it will be easiest to manufacture them inside host animals. This could allow future computer processors to be bred, and on a local basis, rather than being fabricated solely inside multi-billion-dollar semiconductor plants as happens today. Almost certainly, this would raise all kinds of animal rights issues – and even if we managed to breed animals with two or more brain-processors, so enabling the 'product' to be surgically removed without slaughtering the biofactory that created it. Figure 2.1 provides a fanciful illustration of a future biological microprocessor.

If the above does come to pass, then all kinds of ramifications may have to be considered. For example, what would happen if your company's new server started thinking pig-thoughts? Or your smartphone took on the behaviour of the mouse or rat that grew its bioprocessor? OK, so we are now getting into very weird and seriously fantastical territory. Though once we truly learn to engineer biology, these are the kinds of possibilities with which we may have to contend.

FEEDSTOCKS IN THE SKY

Returning to reality, even the widespread adoption of microbial synthetic biology will raise some critical future questions. Not least, there will be the thorny issue of obtaining an adequate supply of organic feedstocks. As Amyris and others

Figure 2.1: A Future Biological Microprocessor?
Image: Christopher Barnatt.

have now proven, in the future we could wean ourselves off petroleum derivatives by switching to synthetically fermented bio-alternatives. However, we do need to remember that we live on a planet with a growing population, and where around 800 million people already do not have enough to eat. Due to climate change and predicted shortages of oil and fresh water, the production of food is going to get harder and harder in the decades ahead. So where will the land come from to grow the billions of tonnes of organic feedstocks that future synthetic biology biofactories may require?

As I have already hinted, one answer could be to feed future colonies of synthetically engineered microbes not on land-intensive crops like corn and sugar beet, but on algae. Microalgaes (or in other words the green stuff that gathers in stagnant water) can be grown in any location in which sunlight is available. Meanwhile parts of our vast ocean surfaces could be used to farm enormous quantities of macroalgaes, otherwise known as seaweed.

Even more radically, rather than raising crops and animals horizontally across billions of acres of land, we could instead cultivate them vertically in urban farms. While this may sound crazy, the idea of 'vertical farming' or 'verticulture' has been gaining momentum for some time, with the first global Vertical Farming and Urban Agriculture (VFUA) conference held in Nottingham University in the United Kingdom in September 2014.

Vertical farms are multi-storey buildings used to grow crops and raise animals. One of the key ideas behind them is to produce food and other biomass within the cities where they will be consumed, so reducing the vast quantities of fuel and other resources that we currently waste on transportation.

With their multiple fields stacked one on top of the other in a controlled environment, vertical farms could allow year-round crop production with no risk of weather-related damage. They should also reduce crop losses in transportation and storage to zero, as well as lowering agricultural water consumption by using hydroponics or aeroponics to hydrate their plants. In fact, if the crops in vertical farms were to be cultivated using aeroponics (which sprays a nutrient-laden mist onto their roots), water consumption compared to conventional farming could be reduced by up to 95 per cent.

A major advocate of vertical farming is Professor Dickson Despommier of Columbia University, who is the author of the groundbreaking book *The Vertical Farm*. As Despommier argues, vertical farms could be used not just 'to reduce the amount of travel between the tomato and your plate', but as a force for urban regeneration. Vertical farms could additionally purify the billions of gallons of sewage and brown water that are created in every city each year, as well as helping to reconnect city dwellers with the natural world. On top of all this, vertical farms may one day cultivate SM plants and animals that would directly fabricate biofuels, biochemicals and final bioproducts.

Of course, growing crops and raising animals in future skyscrapers in the middle of cities will not be without its problems. Not least, you may wonder how plants grown in multistorey buildings would ever get enough light. Yet to this key challenge there are already some answers, including the incorporation of light redirection technologies into buildings, growing plants in slowly rotating stacks, and the use of LED grow lamps. The latter are special LEDs that emit only the 2 per cent of the visible light spectrum used by plants in photosynthesis, and which can hence prove very energy efficient.

If it also worth noting that worldwide there are already a number of operational vertical farms. One of these is The Plant in Chicago, which cultivates both plants and fish in a zero-waste, closed aquaponics loop. What this means in practice is that excrement from the fish is used to fertilize the plants, whose roots and other waste are then fed to the fish.

In Suwon in South Korea, a three-storey vertical farm has been built by the government for research purposes, and is already growing lettuces at record speed under LED grow lamps. Meanwhile in Singapore, a vertical farm called Sky-Greens is managing to profitably supply salad vegetables to supermarkets.

According to Professor Qichang Yang from the Chinese Academy of Agricultural Sciences, China is investing very heavily in vertical farming research and development, with 40 research institutes looking into all aspects of indoor and vertical urban agriculture. With a very large population to feed, China has an understandable ambition to become a world leader in industrial scale urban agriculture, and in turn this could leave the country very well placed to locally cultivate not just food, but also future synthetic biology feedstocks.

MIMICKING LIFE

Vertical farms are set to dominate the skyline of many a metropolis. Their creation should allow synthetic biology to go mainstream, as well as facilitating the emergence of a new, distributed bioeconomy that will make our cities far more self-reliant. Even so, constructing tens of thousands of massive new buildings will not necessarily be good for the environment. For a start, making one tonne of cement results in about a tonne of greenhouse gas emissions. So what if we could create living buildings, or at least buildings coated with materials that would have a positive environmental impact?

Working toward this goal is an extraordinary woman by the name of Rachel Armstrong. In her book *Living Architecture: How Synthetic Biology Can Remake Our Cities and Reshape Our Lives*, she explains how 'all modern buildings are constructed from functionally inert materials that form a barrier between human beings and nature'. To try and address this issue, Armstrong proposes the architectural adoption of 'protocells'. These intriguing molecules arise when oil and an alkaline solution are mixed, and can 'hold a chemical conversation with each other'.

While protocells are not alive, they nevertheless mimic living systems by exhibiting 'a range of lifelike behaviours such as movement, sensitivity and the production of microstructures'. As Armstrong further explains:

> Protocell technology offers a new approach to the engineering and construction of buildings as a fundamental tool for the design of ecological materials. The self-assembly of protocells from a bottom-up perspective can be coupled to the use and recycling of locally available resources that are continuously informed by their surroundings.

What the above means is that, within ten years, protocells may be used to create 'carbon-fixing building coatings'. These would turn our city's carbon emissions into inorganic carbonates, so helping to reduce climate change. At the same time, the carbonates slowly excreted by protocell coatings would further strengthen and insulate building walls, and could even heal the cracks that sometimes open up as buildings age.

Further into the future, surface coatings may be developed with the ability to synthetically fabricate useful substances. If this ever occurs, the outside of every building could become a biofactory capable of fabricating fuels or other chemicals. Alternative protocell building coatings could form a symbiotic foundation for future SM algaes that could be grown vertically as feedstocks for nearby biofactories.

Rachel Armstrong additionally imagines how protocell technologies may one day sustainably reclaim the city of Venice. Here, light-sensitive protocells could travel through the water to seek out the sinking city's woodpile supports. They would next use dissolved carbon dioxide 'to create insoluble crystalline skins from minerals in the water', and which would 'accrete on the woodpiles and gradually petrify them'. In time, an artificial limestone reef would be created by barnacles and other indigenous marine life working in combination with the synthetic protocell technology. The end result could be an expanded underwater formation that would spread Venice's weight distribution, so preventing it from disappearing into the mud.

EMBRACING RADICAL CHANGE

New manufacturing technologies are always most gainfully employed when they are used to make entirely new products, rather than facilitating the production of old things in new ways. As we have seen in this chapter, in the future synthetic biology could be used to fabricate bioplastics and other materials from which we could make everyday items like toothbrushes. In time, we may even learn to engineer plants that will produce a harvest of toothbrushes, as well as micro-organisms that will allow the domestic fermentation of toothpaste. However, if we pursue this kind of pathway, we risk missing the true potential of our extraordinary new science.

Rather than using synthetic biology to help fabricate traditional toothbrushes and toothpaste, it would be far better to engineer a new breed of personal micro-organism that could be cultivated in our bathroom cabinets, and which we could rinse around our mouths every night and morning. This new form of tooth cleaning 'synthetic antibody' would attach itself to plaque, rapidly kill this pesky little variety of human mouth bacteria, and reduce it to chemicals that would harmlessly dissolve in water or saliva. A quick rinse and spit would hence leave our mouths far cleaner than ever possible after countless hours of attention with a conventional toothbrush and toothpaste.

A plaque-destroying personal synthetic antibody may or may not be created. But it will be innovators prepared to adopt this kooky kind of thinking who will become the most successful synthetic biology pioneers. Indeed, once we have learned to routinely program and control life at the component level, we may discover that many of the dumb, branded chemicals that we hoard in our homes are things that we can happily live without.

Just as biology provides the best mechanism for making things out of living cells, so it may also present us with the best tools for shaping a more sustainable society in which we are more in touch with the natural world. This also has to be synthetic biology's greatest potential. Many people may fear the unintended and potentially damn-right-evil consequences of turning biology into just another form of programmable computer data. But as we shall explore further in the next chapter, the digital union of synthetic biology, 3D printing and nanotechnology is likely to herald a quite extraordinary new manufacturing age.

3

NANOTECHNOLOGY 2.0

One of my own, favourite futurists is a former IBM engineer called Thomas Frey. In his book *Communicating With The Future*, Frey introduces the concept of 'attractors', which he describes as imagined future events that shine out as beacons of light in the darkness that lies ahead.

Attractors are created by visionary individuals to provide their followers with a shared destination. As future shaping tools, the most powerful attractors may even prove inspirational enough to change the world.

Probably the most significant attractor ever created was the idea of sending a human being to the Moon. On 25 May 1961, President John F. Kennedy announced this goal in an historic address to the United States Congress. At that time, neither Kennedy nor NASA knew exactly how or even if a lunar landing could be achieved. Nevertheless, the future vision of landing on the Moon galvanized NASA and the American nation to actually make it happen.

About 25 years later, a book called *Engines of Creation* established another seemingly impossible attractor. This highly influential work was written by Eric Drexler, with his grand vision being the development of molecular nanotechnology. Drexler had first put forward his ideas in a 1981 paper published in the *Proceedings of the National Academy of Sciences*. However, it was the release of *Engines of Creation* in 1986 that catapulted the future potential of nanotechnology into the public imagination.

Nanotechnology engineers physical matter on a molecular or even an atomic scale. Technically, the technology operates at a level of precision of between 1 and 100 nanometres, with one nanometre being just one billionth of a metre in size. To provide some context, a human hair is typically around 50,000 nanometres wide, while an average piece of copy paper is roughly 100,000 nanometres thick. Thinking back to the last chapter, an *E. Coli* bacterium is about 3,000 nanometres in length, while a strand of DNA is about 2.5 nanometres in diameter. Depending on the element, individual atoms range from about 0.1 to 0.5 nanometres in size (although down on this scale just exactly what is being measured really comes into question).

TOWARD APM

Nanotechnology may involve a wide range of activities. These range from building things with nanoscale precision using fairly traditional production methods, to creating innovative nanomaterials, to developing 'atomically precise manufacturing' (APM) technologies that craft products on the nanoscale using self-assembly techniques. The latter was very much the focus of *Engines of Creation*, in which Drexler advocated a post-industrial revolution based on APM.

As Drexler laments in the first chapter of his seminal work, despite all of our incredible technological advancements, at present we are still 'forced to handle atoms in unruly herds'. Indeed, if we imagine atoms to be Lego bricks, then all current manufacturing methods require us to assemble things from them while wearing boxing gloves. Granted, since *Engines of Creation* was published we have developed some really cool boxing gloves that allow us a fair level of precision in pushing clumps and piles of atomic Lego around. But picking up and plugging together individual bricks is still a feat beyond the capabilities of any mainstream manufacturing technology.

In *Engines of Creation*, Drexler predicted that one day we will learn to manipulate individual molecules (groups of atoms chemically bonded together), and even individual atoms themselves, in order to make 'nanomachines'. These will in turn manipulate molecules and atoms, and so allow us to fabricate any product we desire. While this may sound like pure science fiction, it is worth remembering that all living things – ourselves included – are self-assembled in this manner on the nanoscale.

Already the pioneers of a new science called protein engineering are learning how to cut-and-paste matter at the molecular level. We can therefore be pretty certain that future nanomachines designed and constructed to human specification are at least a technical possibility. Less radical forms of nanotechnology are also in mainstream application. Indeed, according to the Project for Emerging Nanotechnologies, in 2015 there were over 1,600 consumer products reliant in whole or part on nanotechnology. This figure represents a 24 per cent increase since 2010, and includes microprocessors, computer memory, stain resistant fabrics, self-cleaning glass and water filtration systems. You can learn more at nanotechproject.org.

GRAND INITIATIVES, SMALL SCALE

Since the publication of *Engines of Creation*, a great deal of public money has been invested in nanotechnology. Not least, in 2000 the National Nanotechnology Initiative (NNI) was established in the United States. Since 2001, this has brought together the nanotechnology-related activities of 20 government departments and independent agencies, and is working toward 'a future in which the ability to understand and control matter at the nanoscale leads to a revolution in technology and industry that benefits society'.

Equally recognizing the potential, between 2001 and 2014 over 60 other countries established nanotechnology initiatives. For example, the National Center for Nanoscience & Technology (NCNST) was established in China in December 2003. In March 2006, China also designated nanotechnology research and development as one of the twelve 'mega-projects' included within its *Medium and Long Term Development Plan 2006-2020*.

In 2015, in the United States alone the NNI had a federal budget of over $1.5 billion, and since 2001 has invested nearly $21 billion in nanotechnology development. This makes the NNI the single largest non-military scientific endeavour since the Apollo Moon landings program. It is therefore hardly surprising that, when President Bill Clinton announced the creation of the NNI, he made clear the kinds of radical innovations that it was expected to help deliver. To cite just part of the speech he made on 21 January 2000:

> Imagine the possibilities: materials with ten times the strength of steel and only a small fraction of the weight – shrinking all the information housed at the Library of Congress into a device the size of a sugar cube – detecting cancerous tumors when they are only a few cells in size. Some of our research goals may take 20 or more years to achieve, but that is precisely why there is an important role for the federal government.

While the above goals may once have sounded implausible, today new stronger, lighter nanotech materials have actually been created. Microprocessors and solid state computer storage have also advanced significantly, while medical diagnosis is starting to be transformed. The NNI and parallel initiatives in other countries have been at the forefront of some of these nanotech innovations. Even so, some nanotechnology advocates – including Eric Drexler – believe that the NNI and similar government undertakings have become sidelined from their initial remit.

In part to highlight his concerns, in 2013 Drexler published another book – *Radical Abundance* – in an attempt to get his vision of a nanotechnology revolution back on track. As in his earlier work, the focus in *Radical Abundance* is on the creation of nanomachines capable of atomically precise manufacturing. In contrast, to date the NNI and most other nanotechnology initiatives have focused on the creation of nanostructures using traditional technologies, as well as on the development of new nanomaterials.

As Drexler laments in *Radical Abundance*, since 1986 nanotechnology has come to be defined by size alone, which means that the field has 'broadened to embrace far more than nanomachines and atomically precise fabrication'. As a consequence, scientists have focused on projects with short-term paybacks – such as the production of nanomaterials – rather than the development of atomically-precise manufacturing. In turn, this means that some critical areas of science and engineering have been excluded from major nanotechnology programs. Indeed, as Drexler explains the situation:

> In Washington the promoters of a federal nanotechnology program sold a broad initiative to Congress in 2000 and then promptly redefined its mission to exclude the molecular sciences, the fields that comprise the very core of progress in atomic precision. Thus, the word 'nanotechnology' had been redefined to omit (and in practice, exclude) what matters most to achieving the vision that launched the field.

The above quotation is very important to keep in mind in our exploration of nanotechnology as a Next Big Thing. To be absolutely clear, what Drexler popularized in 1986 was the future creation of 'bottom-up' manufacturing processes that would build products out of individual molecules. The fact that scientists and engineers have since started to deliver the 'top-down' fabrication of nanostructures and nanomaterials is truly remarkable. But we really must not mistake the current reality of top-down nanotechnology as representative of what can and will in the future be achieved. Further, we must not be fooled into believing that Drexler's vision for bottom-up nanoscale manufacturing was wrong because almost all work to date has focused on far less radical, top-down nanotechnology innovations.

THE RISE OF THE MICROFABRICATOR

This chapter is called Nanotechnology 2.0 to remind us that while the first, top-down revolution of Nanotechnology 1.0 is well underway, a second generation of mainstream, bottom-up nanotech has yet to be delivered. This point made, we do now manufacture over one billion nanomachines every year. We just happen to call them 'microprocessors'. These amazing devices process information on the nanoscale, with some of Intel's latest Xeon chips featuring over 5.5 billion tiny transistors.

The modern world has come to depend on microelectronic nanomachines to achieve many of its core economic, social and cultural functions. In this respect, Drexler's vision of a New Industrial Revolution founded on the back of nanomachines has already arrived. But as I hope you appreciate, the true challenge and opportunity ahead is the genesis of a second generation of nanomachines that will manipulate physical matter rather than information. This will be the world of Nanotechnology 2.0 – an age in which atomically precise manufacturing (APM) will allow products to be created on demand not just by 3D printers and synthetic life forms, but in addition by nanomachine 'microfabricators' that will rearrange molecules and even individual atoms to locally assemble whatever we desire. Or as Drexler explains in *Radical Abundance*:

> Where digital electronics deals with patterns of bits, APM deals with patterns of atoms. Where digital electronics relies on nanoscale circuits, APM relies on nanoscale machinery. Where the digital revolution opened the door to a radical abundance of information products, the APM revolution will open the door to a radical abundance of physical products.

Nanotechnology 1.0 may have given us the microprocessor, and with it the Information Age. Yet the computing and Internet revolution of the past few decades may pale into insignificance in comparison to the industrial, social and cultural transformation likely to be catalyzed when microfabricators enter the fray.

NANO FOUNDATIONS

In explaining the distinction between Nanotechnology 1.0 and Nanotechnology 2.0, I may have given the impression that the former is not that important. If so, then I wholeheartedly apologize, as the top-down fabrication of nanostructures and nanomaterials is highly significant and increasingly so. Before we delve into the future self-assembly world of microfabricators, it is therefore worth focusing for a page or few on current and probable-future top-down nanotechnology.

Today the most mainstream nanotech fabrication process is nanolithography. This makes nanoscale patterns on a surface, and is the science and art that facilitates the manufacturer of modern microprocessors, computer memory and other microchips. The technique typically uses ultraviolet light to project images of circuit layers onto a plastic film atop a silicon wafer. Chemical vapours, heat, ion beams and various other materials and methods are then used to 'develop' the exposed images into nanoscale electronic components, insulators and conductive wiring.

In 2015, Intel launched two new generations of microprocessor – called Broadwell and Skylake – that are fabricated using a 14 nanometre production process. These contain components only 14 nanometres apart, with interconnects (wiring) as tiny as 52 nanometres across, and a minimum component feature size of 42 nanometres. A 10 nanometre manufacturing process called Cannonlake is due in 2017, with a 7 and 5 nanometre roadmap already on the cards. This means that Intel is already planning for the sale of mainstream consumer products with some internal components measuring less than 100 atoms wide and spaced at roughly 25 atom intervals.

Also heading in this very tiny direction are IBM, who in July 2015 announced the fabrication of a functional 7 nanometre test chip with partners including GlobalFoundries and Samsung. This advancement is likely to result in the manufacture and sale of chips with as many as 20 billion individual transistors as early as 2017. And all this, you must remember, will be achieved using 'conventional', top-down nanotechnology.

In addition to being used to make silicon chips, nanotechnology is finding a home in the development of next-generation batteries, solar cells and display screens. Often this involves the use of carbon nanotubes (CNTs) or graphene, both of which have a whole host of future applications.

As illustrated in figure 3.1, carbon nanotubes are hexagonal lattices of carbon atoms bonded into tiny tubes a few nanometres in diameter, and usually only a few micrometres (thousands of nanometres) in length. In comparison, graphene consists of a single layer of carbon atoms arranged in a honeycomb fashion. If you are looking at figure 3.1 and wondering if a carbon nanotube could be unrolled into a

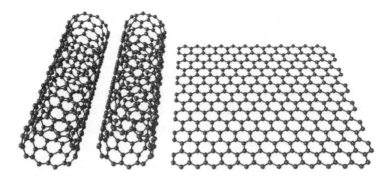

Figure 3.1: Carbon Nanotubes & Graphene.

sliver of graphene, then you are not alone. In fact, back in 2010 at Rice University, pioneers Alexander Sinitskii and James M. Tour managed to use sulphuric acid, potassium permanganate and other chemicals to 'unzip' carbon nanotubes by breaching some of their bonds. This caused their cylindrical structure to unfurl into a graphene ribbon.

Both carbon nanotubes and graphene have similar and extraordinary properties. These include great strength, considerable flexibility and high electrical conductivity. The latter characteristic offers many possibilities to use either carbon nanotubes or graphene to make battery electrodes. Already a team led by Professor Yang Shao-Horn at the Massachusetts Institute of Technology (MIT) has demonstrated how lithium ion rechargeable batteries can be made ten times more efficient by making one of their electrodes out of carbon nanotubes. Meanwhile, back in 2013, Michigan-based XG Sciences launched a new graphene-based electrode material which has the potential to increase lithium ion battery capacity by a factor of four.

In addition to rechargeable batteries, in the future we may also start to power some devices using ultracapacitors. While batteries store electricity electrochemically, ultracapacitors (also known as supercapacitors) maintain an electrostatic field between two plates. In most traditional capacitors, these plates are sheets of metal foil separated by a sheet of paper impregnated with an electrolyte, all wound tightly in a spiral. This construction format allows a reasonably large plate area to be efficiently stored in a relatively small volume, with the plates only a fraction of a millimetre thick and a similar distance apart. But replace the metal foil with graphene, and an ultracapacitor can be created with a larger plate area in a device of the same size.

Researchers at the California NanoSystems Institute (CNSI) have already made graphene-based ultracapacitors with an energy density of 40 watt-hours per kilogram, compared to 28 watt-hours per kilogram for a conventional ultracapacitor. These devices are flexible, fully charge in seconds, and can be made thinner than a sheet of paper. They may therefore be used to make power sources for a wide range of devices, including wearables and electric vehicles.

Graphene may soon also be used to help create highly efficient, flexible photovoltaic solar cells. In fact, in 2012, researchers from Universitat Jaume I in Spain and Oxford University used a combination of titanium oxide and graphene to produce a prototype of such a device.

One day, it may even become possible to print graphene-based solar cells onto any surface, not to mention graphene batteries or ultracapacitors to store their power. This would free us from the constraints of conventional, discrete battery technology. Just imagine clothing, window panes, car bodies and even roads capable of generating and storing electricity.

Graphene additionally has the potential to replace indium-based electrodes in organic light emitting diodes (OLEDs). This will allow low-power, flexible display screens to be created that do not have to incorporate precious metals. Also working to improve future displays, a team led by Stephen Chou at Princeton University has invented a new nanotech structure that they call PlaCSH (and which stands for 'plasmonic cavity with subwavelength hole-array'). This can be incorporated into LEDs and OLED screens to increase their brightness and efficiency by 57 per cent, and screen clarity by 400 per cent. It does this by preventing light from remaining trapped inside the LED, with the nanoscale structure manipulating the light on a scale smaller than a single wavelength.

NANOCOMPOSITES & NANOCOATINGS

In addition to innovating new electronic components, many companies are now developing 'nanocomposites' that are a mixture of a conventional material mixed with a nanoscale additive. Sometimes these nanoscale additives are carbon nanotubes, with these miniscule but very strong strands having already been added to automotive paints, resins, metals and glass in order to make them stronger or more scratch resistant. In the future, materials reinforced with carbon nanotube or graphene additives will allow cars and aeroplanes to be made to the same structural specification as today, but using less material, so saving resources in manufacturing. Vehicles made using nanocomposite materials will also be lighter, and hence more fuel efficient.

Both carbon nanotubes and graphene have also started to be added to plastics to make them stronger and electrically conductive. This development may prove particularly significant in 3D printing, where the ability to extrude a conductive filament raises all kinds of possibilities. Already 3DXTech have released two carbon nanotube composite filaments called 3DXNano ESD ABS and 3DXNano ESD PETG. These are composites of carbon nanotubes mixed with the thermoplastics ABS or PETG, and allow low-cost 3D printers to fabricate parts that offer electrostatic discharge (ESD) protection. In practice this means that the 3DXNano filaments may be used to 3D print parts of hard drives and other electronic components.

In 2015, 3D printing pioneer Graphene 3D Lab launched its Conductive Graphene Filament. This is a nanocomposite of PLA mixed with graphene, and allows the desktop 3D printout of components such as capacitive touch sensors (as used in games controllers and keyboards), as well as circuit interconnects and electromagnetic shielding.

Graphene 3D Lab's ambitions also go far beyond the sale of a single new 3D printing filament. As the company explains in its investor presentation:

> Our goal is to bring to market cutting-edge 3D printing technology that exploits graphene, a material with incredible properties. Our proprietary method has [the] potential to enable a 'one-touch' capability that can print working electronic devices. This as of yet unrealized advancement may become the manufacturing process of choice in nearly every industry.

While carbon nanotubes and graphene tend to get most of the attention, other nanoparticles are also being used to create nanocomposites and nanocoatings. For example, nanoscale particles of titanium dioxide are now routinely applied to glass to make it 'self cleaning'. This works because titanium dioxide is a photocatalyst that generates electrons when struck by UV light, which helps to dislodge dirt. Titanium dioxide is also hydrophilic, which means that it attracts water. This causes rain to spread out evenly across its surface (rather than beading into droplets), before draining evenly down the window a bit like a liquid squeegee. The coating of titanium dioxide applied to self cleaning glass is typically around 25 nanometres thick, which means that it only reduces light transmission by a few per cent.

Several companies have similarly used nanotechnology to produce water-resistant and stain-resistant fabrics. Market leader Nanotex, for example, sells coatings that add hydrophobic 'nano whiskers' to individual fabric fibers. These whiskers are about 10 nanometres long, and repel and elevate liquid droplets, so causing them to bead and roll off the fabric surface.

Another material commonly mixed into nanocomposites or applied as a nano surface coating is silver. Silver nanoparticles have unique optical, electrical and thermal properties, and are hence already being used to help create conductive inks, chemical sensors and organic solar cells. Given that silver nanoparticles are also a powerful antibacterial, they are additionally finding a wide range of applications in hygiene and healthcare products. Silver nanoparticles have even been incorporated into socks and sportswear to kill bacteria and prevent odours.

I could proceed for many more pages to detail a wide range of products – from food packaging to lubricants, nutritional supplements to air filters, and cosmetics to sunscreens – that already incorporate nanocomposites or nanocoatings. But hopefully by now you have got the message that nano additives offer a plethora of improved product opportunities. Virtually no manufacturing industry will be untouched by the widening roll-out of nanocomposites and nanocoatings in the next 10 or 20 years. This is also something to be expected given the amount of public and private money that has been invested in top-down nanotechnology over the past couple of decades.

While the impact of nanocomposites and nanocoatings on certain products and industries is fairly easy to predict, the wider implications are less well understood. For example, while nanocoatings offer manufacturers of glass or fabrics a new product opportunity, for those who clean windows, or who sell detergents, washing machines or laundry services, they could lead to a significant reduction in business. Imagine if most windows, paints and other surfaces never got scratched or dirty – or if most fabrics never ripped, creased, got stained, or smelt bad after days or even months or years of wear.

We have become used to a world in which most products are either disposable or high maintenance, and yet the mainstream application of nanocomposites and nanocoatings could change this situation. In turn, this could lead to significant economic disruption in some service sectors. Nano-enhanced materials will undoubtedly deliver significant benefits. But we must remain alert to the fact that small changes to products on the nanoscale could impact our lives and economies on a very broad scale and not always in a positive fashion.

SELF-ASSEMBLY & APM

While the benefits of nanocomposites and nanocoatings may be considerable, their production basically involves chucking nanoparticles into a mixture or throwing them at a surface. OK, so such a casual dismissal of a whole heap of cutting-edge science and engineering may sound a bit callous. Yet it should remind us that the innovation and application of nanocomposites and nanocoatings has little to do with the development of future atomically precise manufacturing (APM). No significant control exists over where a particular molecule or atom ends up in a nanocomposite or nanocoating. However, if we are to progress toward Nanotechnology 2.0 and the creation of microfabricators, gaining such control will be essential.

Atomically precise fabrication may potentially be achieved in two ways. The first is known as 'positional assembly', and involves the use of conventional-scale hardware to move individual atoms or molecules around. In 1990, an IBM Fellow called Don Eigler became the first person to achieve positional assembly when he used a scanning tunnelling microscope (STM) to move 35 xenon atoms across a flat, crystalline surface. This allowed him to spell out the letters 'IBM', so creating the first nanoscale company logo.

Positional assembly using STM hardware is very impressive indeed. Nevertheless, given that most products contain quintillions of atoms, moving them around individually using any form of top-down technology is incredibly unlikely to develop into a viable manufacturing method. Nanotechnology 2.0 and Drexler's vision of future APM will therefore require a radically-different bottom-up approach. More specifically, we will need to develop nanoscale systems and components that utilize 'self-assembly'. Or as Drexler notes in *Radical Abundance*:

> ... from the start, self-assembly is the line of research that I have advocated as a path toward APM-level technologies. The comparatively meager state of the art in nudging [atoms or molecules] with scanning probe microscopes has been a distraction.

Self-assembly refers to manufacturing processes in which nanoscale parts fit themselves together without the intervention of production tools. This becomes possible when each individual part has a distinct set of bumps and hollows that can catch and lock-on to other components when they are mixed together. The whole idea is akin to putting all of the parts required to make a smartphone into a cocktail shaker, giving it a really good workout, and opening the container to find the latest, fully-assembled Samsung or Apple device.

The above may sound ridiculous. Even so, pioneers of genetic modification and synthetic biology are becoming expert at manufacturing custom DNA chains by building them up from nucleotide molecules using self-assembly methods. They achieve this by bonding the first nucleotide in their desired DNA sequence to a solid surface, before introducing a liquid containing many copies of the second nucleotide. Via natural thermal motion, the molecules of the second nucleotide move around, brushing up against the end of the first nucleotide, and when the two come together just right the parts bond. Leftover material is then washed away, with chemicals added to prepare the end of the second nucleotide to accept the third element in the sequence. This 'blind assembly process' is repeated over and over as each subsequent nucleotide is added.

As the above hopefully makes clear, it is possible to self-assemble individual molecules into longer polymer chains using no nanoscale machinery whatsoever. As may be expected, the longer the chain being built, the more likely errors will occur, and so bad chains do need to be discarded. To help join short chains of molecules into evermore complex structures, biological engineers can also turn to bacterial molecular machines known as enzymes.

Enzymes are catalysts for biological reactions that can be used to cut, paste and transcribe polymer chains. For example, 'restriction enzymes' can brush up against proteins or DNA molecules to 'read them by touch'. They then adhere in just the right place, before making a cut by rearranging appropriate atoms. Other enzymes can similarly match-and-splice the ends of two molecular chains together.

As we saw in the last chapter, using the above methods practitioners of synthetic biology have already built novel strains of bacteria and yeast as a new form of organic production technology. In the future, it may even be possible to go further in order to use protein molecules as the foundation for atomically precise manufacturing.

PROTEIN ENGINEERING & BEYOND

We already understand how some protein molecules serve basic mechanical functions. Within human muscle, for example, there are proteins that push or pull, or which act as struts or bearings. Bacteria even propel themselves along using corkscrew protein propellers attached to reversible, variable-speed protein motors. Proteins therefore present engineers with a toolbox of potentially very useful nanoscale mechanical components.

By mixing-and-matching the appropriate proteins, we may one day learn to construct complex mechanical machines on the nanoscale using self assembly methods. Scientists have in fact already managed to manufacture sophisticated biological machines – such as viruses – by mixing together the appropriate proteins and other chemicals, with self-assembly causing the right parts to snap together in the right place.

A particularly promising development was reported in September 2015 by researchers from Queen Mary University of London (QMUL). Here a technique has been developed for creating complex tubular tissue structures using proteins and amino acid compounds called peptides that are mixed together in a solution. When these organic, chemical components touch each other, they 'self-assemble to form organic tissue at the point where they meet'. This can allow the controlled creation of complex shapes via 'natural' processes that include growth and healing. The research could lead to the artificial, self-assembly of tissues including veins and arteries. Or as Alvaro Mata, Director of the Institute of Bioengineering at QMUL enthused:

> What is most exciting about this discovery is the possibility for us to use peptides and proteins as building-blocks of materials with the capacity to controllably grow or change shape, solely by self-assembly.

Scientists are now also learning how to design artificial molecular chains called 'foldamers'. These mimic the ability of natural proteins, and fold into pre-determined configurations a bit like nanoscale Lego blocks with pre-configured bumps and hollows. One day, such foldamers may be able to self-assemble into nanoscale machines that will in turn be able to fabricate larger structures. Just remember the idea of making a smartphone by mixing the right parts together in a cocktail shaker. All that we need to make this work is to learn how to design and chemically manufacture the correct molecular parts that will naturally lock into place when blended in the right combination.

These kind of developments will sadly take decades rather than years to become a common reality. There is, after all, an immense developmental chasm between mixing together proteins and peptides to make a synthetic virus or tissue structure, and the production of microfabricators able to arrange molecules entirely to human whim. Proteins also have shortcomings as engineering modules, including being highly intolerant of hot and cold temperatures. We do not currently build many things out of fragile biological materials for good reason. Yet just as our recent ancestors used no more than their flesh-based bodies to craft things out of metal and stone, so future nanomachines made from synthetic protein-like polymers may be able to fabricate products out of bioplastics, metals and highly resilient nanomaterials including graphene and carbon nanotubes.

Natural systems currently build in nature's own biological image. Yet synthetic biology remains free to evolve self-assembly manufacturing methods on a far wider basis. Or as Eric Drexler so powerfully explains in *Engines of Creation*:

> . . . second generation nanomachines will do all that proteins can do, and more. In particular, some will serve as improved devices for assembling molecular structures. Able to tolerate acid or vacuum, freezing or baking, depending on design, enzyme-like second-generation machines will be able to use as

'tools' almost any of the reactive molecules used by chemists, but will wield them with the precision of programmed machines. They will be able to bond atoms together in virtually any stable pattern, adding a few at a time to the surface of a workpiece until a complex structure is complete. Think of such nanomachines as assemblers.

Because assemblers will let us place atoms in almost any reasonable arrangement... they will let us build almost anything that the laws of nature allow to exist. In particular, they will let us build almost anything we can design, including more assemblers. The consequences of this will be profound, because our crude tools have let us explore only a small part of the range of possibilities that natural law permits. Assemblers will open a world of new technologies.

EXPANDING POSSIBILITIES

Alongside Eric Drexler, another nanotechnology advocate and guru is Ralph Merkle. In a 2009 lecture for the Singularity University, Merkle presented a simple but extremely powerful Venn diagram to highlight the manufacturing possibilities that lie ahead. A version of this graphic is reproduced in figure 3.2, with a small dot representing all of the products that we can make today. The large circle surrounding the dot then represents all of the physical things (arrangements of atoms) that could potentially exist, and which future nanotechnology should allow us to actually fabricate. The diagram hence highlights 'how far we are from a general ability to build things', and in turn the extraordinary potential of nanotechnology.

So just what are the amazing, future things that nanotechnology will allow us to manufacture? Well, as I

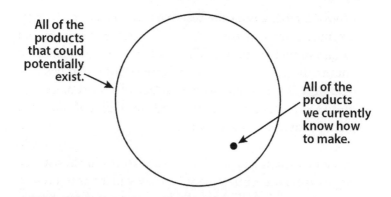

Figure 3.2: The Potential of Nanotechnology.
Adapted from Ralph Merkle, 2009.

have already mentioned in this chapter, soon even 'basic', top-down nanotech will allow us to fabricate next-generation microprocessors with tens of millions of transistors per chip. In fact, early next decade, we will start to manufacture microprocessors with *hundreds of billions* of transistors per chip – or in other words, more transistors per chip than there are neurons in the human brain.

Future nanotechnology will also assist significantly in our battle to overcome resource scarcity. As we have already seen, new nanomaterials will allow vehicles to become lighter and hence more energy efficient. Yet far more radically – and as we shall explore in Part III of this book – nanotechnology will also help us to obtain fresh energy and raw materials from space.

Along with bioprinting and post-genomic medicine, nanotechnology is additionally destined to significantly improve human health, not to mention assisting our evolution into a transhuman species. Does this mean that, 20 years from now, many people will have artificial nano-technologies incorporated inside their bodies? Well yes, as we shall see in chapter 10, absolutely it does.

Alongside all of these developments, nanotechnology may even help us to solve those grand challenges that stand in the face of our long-term survival. For example, nanotechnology may allow the creation of macro-scale greenhouse gas filtration plants. These would remove climate-change-inducing gases from the atmosphere, and could potentially turn the resultant waste carbon into something useful like graphene or carbon nanotubes. This said, when it comes to solving grand challenges and having a fundamental impact on our lives, the greatest potential for nanotechnology has to be the creation of future multi-purpose 'nanofactories' or 'microfabricators'.

In theory, future nanotech manufacturing hardware will be able to take any suitable feedstock and self-assemble it into any product we desire. It will also do this on a local basis, and ought to be able to recycle one product into another. The technology to do this is unlikely to arrive until at least 2035. But as today's pioneering work in protein engineering already demonstrates, nanofactories are at least a theoretical, technical possibility.

So what may the future of manufacturing actually be like? Well, as you may guess, probably the best vision to date has been provided by Eric Drexler. As he outlines in *Radical Abundance*, we should expect enormous automobile factories full of multi-million dollar equipment to be replaced with 'garage-sized' facilities that 'assemble cars from inexpensive, microscopic parts, with production times measured in minutes'. Such future nanofactories would take the principles of modern automated manufacturing, scale it down, add knowledge gained from processes like current microchip manufacture, and add in biomolecular systems engineering and organic chemistry in order to allow very large, complex items like vehicles to be manufactured on demand.

To help us imagine how all of this would work, Drexler invites us to picture a future garage-sized nanofactory with observation windows. If we looked inside, we would see familiar-looking industrial robot arms assembling conventional-scale car components. But look to one side, and we would also see a wall of smaller chambers in which the traditional-scale car components are first being manufactured by smaller robot arms. In turn, the parts and materials being used to fabricate the traditional components would be being fabricated by even smaller robot arms within even smaller chambers. As Drexler goes on to explain:

> ... with the aid of progressively greater magnification you [would] see a sequence of smaller and smaller machines and chambers leading ultimately back to machines that build components by assembling atomically precise microscale building blocks.

Over the past few decades, many works of science fiction have imagined bottom-up nanofabrication systems wherein clouds of nanobots create products almost out of thin air. In contrast, Drexler's vision of a nanofactory in which millions of robots progressively make parts for assembly by robots bigger than themselves is a far more realistic future vision.

As Drexler's books explain in detail, molecular-level fabrication could be achieved by nanoscale robotic arms and other mechanical systems operating at the very high speeds possible on the nanoscale. It would therefore be possible for the machines labouring in the nanoscale and microscale chambers in a future nanofactory to keep up with the supply of conventional-scale components required by the largest assembly hardware. In theory, Drexler's nanofactories would require no more than a supply of basic atomic feedstocks in order to very rapidly manufacture large products like automobiles, spacecraft, and the humanoid robots that we will turn our attention to in chapter 5.

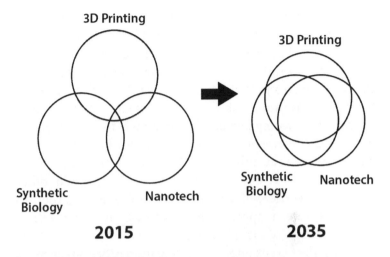

Figure 3.3: Local Digital Manufacturing Convergence.

TECHNOLOGIES CONVERGE

As we have seen in this chapter, nanotechnology will offer some incredible future possibilities. But it will not arrive in isolation. Future nanofactories will one day make products on the desktop, in a garage, or in shipping-container-sized modules. Even so, I very strongly suspect that the technologies within most future microfabrication devices will be based on a mix of 3D printing, synthetic biology and nanotechnology in combinations appropriate to each individual production context and situation.

Figure 3.3 presents a Venn diagram to illustrate the likely convergence of 3D printing, synthetic biology and nanotechnology over the next 20 years. By now this critical concept is hopefully firmly embedded in the amazing, biological computing mechanism that resides in your head. But just to be certain, and before I bring this chapter to a close, let us review how our three newest manufacturing methods are likely to dramatically converge.

As we saw in chapter 1, 3D printing refers to a broad range of technologies that turn digital models into physical things by additively building them up in layers. As we investigated in chapter 2, synthetic biology achieves a similar feat, save that the digital models are stored in DNA code, with the involved fabrication making use of the extraordinary ability of radically re-engineered living things to self-assemble without the intervention of production tools. Finally, as we have explored in this chapter, future bottom-up nano-technologies will engineer proteins, foldamers and other chemistry-centric self-assembly technologies in order to once again turn appropriate feedstocks into functional, physical copies of digital designs.

Today, the best example of a fabrication process that sits at the crossroads of 3D printing, synthetic biology and nano-technology is bioprinting. As we saw in chapter 1, this is already being used to 3D print human tissues for drug testing purposes, and in the decades ahead is likely to allow the fabrication of tissues and organs for human transplant. As you may recall, the basic, bioprinting process involves the 3D printout of layer-upon-layer of a cell aggregate. After printout, the cells that constitute these layers are then left to rearrange and fuse into the far more complex structures that comprise living tissue.

What the above recap signals is how, in its post-printout maturation phase, bioprinting already relies on bottom-up self-assembly to rearrange and fuse the cells that it initially 3D prints. To be absolutely clear, what this means is that current bioprinters already make things via a two-stage fabrication process that first prints out layers of cells, and then takes advantage of their ability to self-assemble.

The printout of living cells that subsequently self-assemble into far more complex structures offers incredible possibilities for the future fabrication of replacement human tissues and potentially also food. In addition, the bioprinting of new synthetic biology feedstocks may facilitate the future microfabrication of all manner of things from living, semi-living and dead-but-once-living materials. Today we grow trees to turn them into houses. But tomorrow we may bioprint buildings from synthetic cells that will continue to live and grow in harmony with our environment.

Away from the organic sciences, the interface of 3D printing and nanotechnology is also receiving much practical attention. Not least, a 'nanophotonic' 3D printing method called two-photon polymerization (2PP) is being developed by several research teams. These include the Additive Manufacturing Technologies (AMT) group led by Jürgen Stampfl at the Technical University of Vienna, and Nanoscribe in Germany, a spin-off from the Karlsruhe Institute of Technology (KIT).

2PP is a vat photopolymerization process that uses a 'femtosecond pulsed laser' to selectively solidify a special photopolymer resin. 2PP 3D printers have already achieved a layer thickness and an X-Y resolution of between 100 and 200 nanometres, which means that they can be classed as nanotechnology. Figure 3.4 shows a 2PP printout of Vienna Cathedral produced by the team from the Technical University of Vienna, with this amazing model being just 0.1 millimetres in length.

In the future, 2PP may enable the very precise fabrication of very tiny things – such as microelectronic and optoelectronic circuits – as well as the rapid manufacture of larger objects. In particular, 2PP may allow future 3D printed objects to be both light and strong, as they could be made not from solids, but from nanoscale lattices that are largely empty space.

As the developments in DNA synthesis and protein engineering outlined earlier in this chapter serve to highlight, synthetic biology and self-assembly nanotechnology are two new sciences that are very much evolving hand-in-

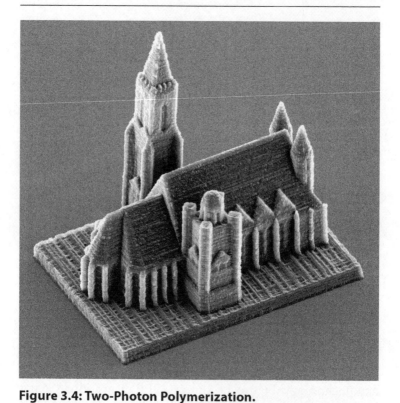

Figure 3.4: Two-Photon Polymerization.
This printout of St. Stephen's cathedral in Vienna was created by Jürgen Stampfl's AMT group at the Technical University of Vienna and is only 0.1 mm in length. Image reproduced with the permission of Robert Liska.

hand. Some research teams are also now integrating 3D printing into the mix. For example, at the University of Illinois a team led by Martin D. Burke is developing a molecular 3D printer that can fabricate carbon-based small molecules 'at the click of a mouse'. Their incredible hardware assembles the molecules in an additive fashion from simple chemical building blocks. The team have already managed to build 14 different classes of small molecules, with possible applications existing in the development of LEDs, solar

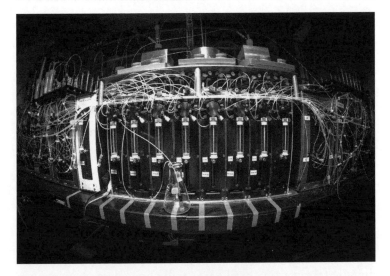

Figure 3.5: A Molecule-Making Machine.
Created by a team lead by Martin D. Burke at the University of Illinois. Photo: L. Brian Stauffer. Image courtesy of BurkeScience / University of Illinois.

cells and medications. Figure 3.5 illustrates their prototype molecule-making machine.

LDM & THE MICROFABRICATOR

We already live in a world in which all forms of information can be digitally created, stored, processed and communicated anytime, anyplace and anywhere. The ubiquitous and low-cost technology that allows this to happen is based on 'microprocessor' devices that manipulate information on the nanoscale. Not that long ago, this extraordinary technology was the Next Big Thing. But, as I argued in the *Preface*, the revolution of computing and the Internet has now come to an end.

In aggregate, 3D printing, synthetic biology and nanotechnology will allow the creation of 'microfabricators' that will facilitate widespread local digital manufacturing (LDM).

In other words, future microfabricators will transform the world of physical object manufacture, storage and exchange in just the same way that the Internet has already transformed the world of information. Already $299 3D printers can produce nanocomposite objects on the desktop, while biofuels and bioplastics can be locally fermented from a range of agricultural feedstocks. And we have barely got started yet.

In comparison to future microfabricators, today's 3D printers, synthetic biology infrastructure and protein engineering tools are more like vacuum tubes than microprocessors. But just as vacuum tubes fairly rapidly evolved into computer chips with billions of times more functionality, so it is very reasonable to predict that today's primitive digital fabrication hardware will one day evolve into microfabricators capable of turning a digital design into a complex, physical object anytime, anyplace and anywhere. Exactly when this will happen is obviously difficult to pinpoint. Though my guess is that general purpose microfabricators will start to emerge somewhere between 2035 and 2050.

A world in which local digital manufacturing is a common reality will be radically different from the world of today. For a start, download-and-fabricate will have replaced manufacture-and-ship, while most products will be customized, and many things will be manufactured by and from living matter. Current attempts to control illegal music or video file sharing will also have been eclipsed by the mass exchange and counterfeiting of digital products. Monetary systems based on the use of physical tokens, like coins, notes or credit cards, will also have been long abandoned in a world in which any physical object will be able to be easily replicated.

As microfabrication technologies become more and more sophisticated from around 2025 onwards, some or all governments are likely to attempt to control what their citizens can and cannot fabricate. And yet in such endeavours I suspect they will fail. By putting the ability to manufacture into the hands of the majority, microfabricators will greatly empower the future population, and the freedom to manufacture is something that I suspect many will fight vigorously to maintain.

The Brave New World of local digital manufacturing will undoubtedly be economically, socially and culturally transformative, and in both positive and negative ways. Yet my guess is that the positives will outweigh the negatives, with crowdsourcing communities sharing objects online, and local economies brought back to life by the opportunities for the majority of people's possessions to produced within a few score miles of their home. This is, after all, a state of affairs that existed for most of human history, and one which could dramatically curtail the catastrophic impact of human civilization on the natural ecosystems that keep us alive.

As all forms of manufacturing and medicine converge on the nanoscale, so knowledge and skillsets will blur across currently diverse industries and industrial sectors. Just as the race-to-binary of the first digital revolution drove all forms of media into cyberspace, so the second digital revolution of local digital manufacturing will lead most production engineers, repair technicians, doctors and food producers toward a reliance on a common set of tools for programming matter on the nanoscale. As this occurs, the volume and complexity of information processing on which our civilization will depend will become mind blowing. So mind blowing, in fact, that no human being will be able to process, let alone comprehend, the digital deluge that will become the common foundation for our everyday survival. The pending age of local digital manufacturing will therefore require the development of sophisticated artificial intelligences able to cope with the practicalities of nanoscale and synthetic biological fabrication. It is therefore to the arrival of this new species that we will turn our attention in the next part of this book.

PART II

SYNTHETIC CITIZENS

4
ARTIFICIAL INTELLIGENCE

In December 2014 Professor Stephen Hawking told the BBC that 'the development of full artificial intelligence could spell the end of the human race'. As the highly respected scientist went on to predict, 'once humans develop artificial intelligence, it [will] take off on its own, and re-design itself at an ever increasing rate'. Human beings 'who are limited by slow, biological evolution' will therefore not be able to compete and will 'be superseded'.

Back in October 2014, similar concerns were expressed by Elon Musk, the founding CEO of both SpaceX and Tesla Motors. In a speech delivered at MIT, Musk pronounced that creating AI is akin to 'summoning the demon', and constitutes one of the greatest threats facing the world today. In January 2015, Bill Gates also expressed similar concerns, stating that he is 'in the camp that is concerned about super intelligence'.

Given the ability and profile of Hawking, Musk and Gates, their anxiety regarding the future of artificial intelligence (AI) should not be ignored. Even so, I do not think that we should be quite so pessimistic. In the decades ahead, the development of AI is definitely going to raise a lot of concerns, and not least as very smart technologies take on jobs once performed solely by human beings. Many people will find such 'mental automation' somewhat scary and downright wrong. Yet machines have been 'replacing' their human creators since the Industrial Revolution, with the overall impact on the majority of the human race generally being very positive. I doubt that many people would want to wind the clock back to the days before powered transport, automated production systems and information technology.

Our civilization has always been founded on intelligence, and so the more we create, the better things ought to become. In fact, the development of AI could prove essential for our very survival. As we saw in the last three chapters, future local digital manufacturing technologies have the potential to help us deal with the looming threat of resource scarcity. We will, however, only be able to master the combined art of 3D printing, synthetic biology and Nanotechnology 2.0 with the assistance of machines far smarter than ourselves. The practicalities of fabricating organic and inorganic objects on the nanoscale are simply too complex to be comprehended, let alone processed, by the current human brain.

In addition to helping with the grand challenge of resource scarcity, AI looks set to deliver major medical advancements. Stephen Hawking may caution that humans 'are limited by slow, biological evolution'. Yet I strongly suspect that future AI will help *Homo sapiens* to escape the limits of natural biology. Unlike many pundits and science fiction authors, I therefore believe that AI will play a key role in advancing rather than repressing the human condition.

This chapter examines the development of those smart and intelligent technologies that are most likely to assist with the survival and advancement of our civilization. Specifically, the following pages will examine how AI will lead to increased mental automation, the development of a new customer interface, the mining of 'Big Data', the improvement of medical diagnosis, the removal of language barriers, and the enhancement of many other new technologies.

THE IMITATION GAME

In 1950 a seminal scientific paper was published by Alan Turing, the father of modern computing. Turing's paper was called *Computing Machinery And Intelligence*, and asked the apparently simple question of whether or not it was possible for an artificial machine to think. To address the conundrum, Turing proposed an 'imitation game' in which an interrogator would ask questions of two hidden subjects, one male and one female. Communication between the interrogator and the subjects was to take place via a teleprinter communicating device. Via text-based communications alone, the interrogator would then have to determine which of the subjects was the man and which was the woman.

With his concept in play, Turing next pondered what would happen if a machine replaced one of the subjects in the game. A machine capable of taking on this role without the interrogator suspecting would subsequently be deemed capable of 'thinking'.

For many years, the challenge of imitating a human being in a text-based communications exchange was taken as the benchmark of machine intelligence. In June 2014, the 'Turing Test' was even deemed to have been passed in a competition held at the University of Reading. The winning computer program was called Eugene, and managed to convince 33 per cent of judges that it was human in a series of simultaneous, five-minute keyboard conversations.

Eugene simulates a 13 year old boy, and was developed over many years by Russian programmers Vladimir Veselov and Eugene Demchenko. Without doubt, the creation of Eugene will stand as an AI milestone. Even so, I think that we should seriously question why any measure of 'artificial intelligence' needs to be based on the ability to imitate a human.

It is widely accepted that dolphins are the second most intelligent species on the planet. Dolphins do, however, live in water rather than air, have no arms or legs, and hence have little habitually or physically in common with a person. We therefore do not even try to assess dolphin intelligence by comparing their mental capability with that of ourselves, let alone by testing if a dolphin can mimic human communication.

The above logic ought equally to apply in any sensible appraisal of AI. At least at first, true AIs are likely to be electrically powered, as well as being manufactured and programmed rather than organically grown. Future AIs will also be very unlikely to eat or sleep, and may well have no independent, physical body. We therefore should not expect AIs to 'think' in a remotely human manner, let alone to have their 'intelligence' measured comparative to ourselves.

Even Alan Turing conceded that his test could be criticised because 'the odds are weighted too heavily against the machine'. As he wrote in his 1950 paper:

> If the man were to try and pretend to be the machine he would clearly make a very poor showing. He would be given away at once by slowness and inaccuracy in arithmetic. May not machines carry out something which ought to be described as thinking but which is very different from what a man does?

As Turing's wise words serve to remind us, defining just what is meant by the term 'artificial intelligence' has always been problematic. Nevertheless, a definition is still rather handy, and herein I will adopt that provided by the Association for the Advancement of Artificial Intelligence (AAAI). This defines AI as an 'understanding of the mechanisms underlying thought and intelligent behaviour and their embodiment in machines'.

The above is both a practical and a useful definition, and not least because it does not seek to benchmark AI against the intelligence or behaviour of a human being. In addition to being Turing-Test neutral, the definition of artificial intelligence provided by the AAAI also comfortably accommodates forms of AI that are either 'broad' or 'narrow'.

MENTAL AUTOMATION

Future broad forms of AI – often referred to as 'artificial general intelligences' or 'AGIs' – will be able to perform any mental task, and could hence supersede humans in a great many occupations. Exactly when AGIs will be created is very difficult to predict, although Ray Kurzweil, Google's Director of Engineering, expects them to exist by 2029. Meanwhile UK cybernetics guru Professor Kevin Warwick is more conservative, placing the arrival of AGIs around 2050. Personally I suspect that Warwick is more likely than Kurzweil to be right on this one. Though I do expect very *smart* and highly *autonomous* technology to be sharing the planet with us by the early 2030s.

While broad AGIs may not yet exist, narrow AIs that can exhibit apparently intelligent behaviour in a predefined field are rapidly advancing. Such AIs will increasingly be able to automate jobs that are currently undertaken by humans, such as answering basic queries in a call centre. The development of 'mental automation' technology now also dates back over 70 years.

During WWII, Alan Turing and his colleagues at Bletchley Park in the United Kingdom created a computer system called Colossus that was capable of deciphering coded German messages. In 1951 at the University of Manchester, a computer was then programmed to play checkers and eventually chess. Over time, such narrow forms of AI have become very sophisticated, with world chess champion Garry Kasparov beaten by IBM's Deep Blue AI way back in 1997.

Today, we increasingly rely on narrow AI systems to accomplish a range of tasks that include autopiloting aircraft, detecting credit card fraud, optimizing the operation of electricity power grids, and performing stock market trades. Narrow AI algorithms are also routinely used to work out the optimal placement of the billions of tiny electronic components and wiring interconnects found inside the latest microprocessors. Such a task can only be undertaken by a computer program, as it is far beyond the mental capacity of any single human being, or even a large team thereof. The development of the microfabricators discussed in the last chapter is similarly likely to depend on the evolution of increasingly sophisticated AIs. It is simply not possible for a human being to comprehend the placement and interrelation of every cell or molecule in even a very small object. But some future AIs will become adept at this task.

THE NEXT INTERFACE

In addition to enabling the automation of both simple and extremely complex mental activities, over the next few years AI will provide a new computing and customer interface. This will happen as people increasingly come to rely on PC or smartphone-based 'virtual assistants' (VAs). At the time of writing, the first useful versions of such programs are just about entering the mainstream, with some people starting to regularly converse with a desktop or pocket device, or even a watch or other wearable.

Occupying this marketplace, Apple's Siri VA has been part of iOS since 2011, while the Cortana VA from Microsoft launched for the Windows Phone operating system in 2014. Cortana additionally launched for the desktop as part of Windows 10 in July 2015, while also becoming available for the Android platform in the same year.

Both Siri and Cortana provide a voice interface that allows users to dictate text, ask questions, set appointments, and perform other potentially useful tasks. Google has a similar offering called Google Now, while in August 2015 Facebook began to rollout a VA called 'M'. All of these AI applications make use of the Internet to learn from the activities of many users. In turn, this allows them to crowdsource their broadening knowledge and cognitive development. How do you build the best virtual assistant? Well, the easiest way is to get a few hundred million people to teach it how to behave.

As all major operating systems become free products, Apple, Microsoft, Google and others may well focus on the development and sale of VAs as one of their most lucrative primary offerings. Granted, VAs are currently being supplied for free in order to build market share and to obtain user feedback. But fairly soon they will be good enough to charge for. That, or VAs will be funded by an integration with paid digital media content, advertising, or the facilitation of e-commerce.

As VAs improve, it is quite likely that people will start to rely on them to accomplish a wide range of digital activities. As a consequence, VAs may well become the most significant new customer interface since the birth of the world-wide web. Just imagine the commercial, social and cultural power that we will vest in VAs if we trust them to undertake even a reasonable minority of our searches, shopping, financial transactions and other online activities. It really is not surprising that, at the launch of Windows 10, Microsoft CEO Satya Nadella referred to the Cortana VA as 'the third platform' in computing (beyond the first two platforms of operating systems and the Internet). Given the market potential, it is quite possible that the next, as-yet-unknown technology giant will be a VA manufacturer.

COGNITIVE COMPUTING & BIG DATA

A level up from AI systems that apply programmatic rules or simple learning algorithms, very sophisticated AIs are starting to be created that can turn their processor cores to a broadening range of mental activities. Most notably, IBM has created 'Watson' as a 'cognitive computing system' that is able to mirror human learning processes. In February 2011, Watson famously managed to beat two human champions on the US TV game show *Jeopardy*. The three-night challenge took place in IBM's T.J. Watson Research Laboratory, where Watson applied its 2,800 processor cores, and a data bank of 200 million pages of text, in order to process the meaning behind those linguistically complex answers to which *Jeopardy* contestants have to provide the questions.

In January 2014 IBM announced the creation of its Watson Group as a 'new business unit dedicated to the development and commercialization of cloud-delivered cognitive innovations'. As the company went on to explain, its investment of more than $1bn into the Watson Group:

> . . . signifies a strategic shift by IBM to accelerate into the marketplace a new class of software, services and apps that think, improve by learning, and discover answers and insights to complex questions from massive amounts of Big Data.

Big Data generates value from the storage and processing of very large quantities of digital information that cannot be analyzed with traditional computing techniques. Until quite recently, very-large-scale data processing was impossible. However, with the growth of computing power, many companies and public bodies are starting to focus their attention on Big Data. In turn, this is driving the development and application of next-generation AIs as the technology that will allow meaning to be extracted from Big Data sets.

Most commentators link the opportunities and challenges of Big Data to the 'three Vs' of volume, velocity and variety. Clearly 'volume' is the defining characteristic of Big Data, with the quantity of digital information created now growing exponentially and reaching extraordinary proportions.

Indeed, according to Cisco, by the end of 2018 global Internet traffic alone will reach 6.5 zettabytes (6.5 billion terabytes) a year – up from 3.1 zettabytes in 2013. Alongside the explosion of digital media, the growth in Big Data volumes is also being driven by the development of the Internet of Things (IoT), with an increasing number of commercial devices and domestic appliances starting to generate their own data stream.

No human being will ever be able to make sense of the ocean of digital information that humanity now chooses to create. But crunching petabytes of data in order to spot patterns and to extract the underlying 'sentiment' will be no problem for future AIs. Just as significantly, AI software will increasingly be able to make sense of those high volume data types – such as genomic information – that are starting to be manipulated and even fabricated from scratch by practitioners of synthetic biology and genetic medicine. It really must be appreciated that all of the developments outlined in Part I and Part IV of this book will rely very heavily on future AI systems capable of processing Big Data.

Big Data 'velocity' also raises a number of key challenges, with the rate at which data is flowing into many organizations now exceeding the capacity of their IT systems. Future AIs that can determine just what data needs to be retained and processed, and what can be safely discarded, will therefore have an important role to play.

The above means that, increasingly, human beings will only be presented with digital information after it has been filtered for relevance and value by one or more AI systems. On the face of it, this may sound like a very scary concept. However, I think that most people will be very grateful to have a VA that reads and *intelligently* pre-filters their e-mails. Such systems are likely to be available in the early 2020s, and could also screen phone calls to stop us being disturbed by marketing spam.

The third key aspect of Big Data is that of increasing 'variety', and once again presents significant opportunities for AI application. Only a couple of decades ago, almost all computer data was generated by human fingers striking keyboards. A fair proportion of such data was also highly structured, with computers largely tasked with the processing of rigid-format text documents, financial transactions, stock records and similar data categories. In contrast today, most digital information is produced by cameras, microphones and other sensors, with at least 80 per cent of data being non-structured. Examples of non-structured data formats include video files, photographs, medical scans and most social media posts, all of which are pretty much impossible for traditional computer systems to comprehend.

As we shall shortly see, developments in AI are starting to allow more and more forms of unstructured data – and in particular digital media files – to be meaningfully interpreted by computers. AI systems are also starting to learn how to extract meaning from unstructured text-based documents. IBM's Watson, for example, is capable of adopting the human learning process of observing, interpreting and analyzing data in order to expose patterns and insights and to make informed decisions. This allows Watson to 'achieve mastery over a given subject and develop expertise'. Or as IBM further explain:

> . . . whereas conventional computing systems are programmed based on rules and logic and follow a rigid decision tree approach, Watson is different. Watson takes in data from all sorts of sources, from research reports to tweets. All the information humans produce for other humans to consume. But importantly, Watson is not bound by volume or memory [as] one instance of Watson can consume 30,000 documents a day.

ENHANCING INSIGHT

Even today, some of IBM's clients are using Watson to 'bridge the gaps in human knowledge' and 'gain better insights'. For example, Deakin University, ANZ Bank and IBM's own technical support are using Watson Engagement Advisor to automate and improve customer interactions.

The Watson Engagement Advisor speaks to customers in natural language, and mines vast data repositories to offer educated, evidence-based answers (rather than standard FAQ responses). The system also learns from each customer interaction, hence allowing it to constantly improve service levels and so reduce the number of customer service requests. Meanwhile the Watson Discovery Advisor is starting to be adopted in companies that want to 'accelerate breakthroughs' by making connections and drawing relationships between data sets. As just one example, IBM is working with *Bon Appétit* magazine to develop a Watson-based 'cognitive cooking system' that can create new recipes.

IBM now also offers a product called Watson Analytics that they claim will 'bring sophisticated Big Data analysis to the average business user'. Watson Analytics is cloud based – so allowing any company to use it as a service over the Internet – and is intended to do 'all of the heavy lifting related to Big Data'. In practical terms, this means that Watson Analytics is able to retrieve appropriate data sets, analyze them, clean up the output, and provide a collaborative environment for communicating the results.

As Ron Miller recently argued on *Tech Crunch*, many companies understand that they should be making better decisions by using Big Data analytics, but they are not doing so because it is too hard. The development of Watson Analytics and similar cloud AI services is hence destined to bring Big Data to the masses.

Already many government agencies and companies (including Amazon Web Services) offer Big Data sets that businesses can correlate with internal information to generate significant insights. Sadly today, many managers believe that Big Data analysis is all about looking more effectively at internal company information, with the benefits of adding AI into the mix therefore being limited. And yet already the potential exists to use cloud AI services to correlate internal company information with public Big Data sets. This can allow a business to find out, for example, exactly how changes in the climate, the economy, or social media sentiment, actually impact its sales and operations. Ten years from now, large companies that continue to 'manage in isolation' by failing to correlate their internal data with external Big Data sources may be unlikely to survive.

Reflecting the above, many analysts believe that AI is reaching a tipping point in terms of business value added. In March 2015, the *Harvard Business Review* carried an article to this effect in which it noted how AI is rapidly becoming 'an important strategic accelerator'. Featured in the article was an interview with venture capitalist Mark Gorenberg, who specializes in investing in analytics and data-centric companies. As he explained:

> AI historically was not ingrained in the technology structure [of a business]. Now we're able to build on top of ideas and infrastructure that didn't exist before. We've gone through the change of Big Data. Now we're adding machine learning. AI is not the be-all and end-all; it's an embedded technology. It's like taking an application and putting a brain into it, using machine learning. It's the use of cognitive computing as part of an application.

DATA DRIVEN MEDICINE

As well as improving analytics and customer engagement in data-intensive industries like financial services, AI is going to have a revolutionary impact in the medical sector. For a start, diagnosis will be improved, with IBM having already partnered with 13 leading cancer institutes in order to develop Watson for Oncology. This will enable more personalized treatments, with the time taken to 'translate DNA insights, understand a person's genetic profile and gather relevant information from medical literature' reduced from weeks to minutes. As IBM further explain:

> Oncologists like all clinicians are struggling to keep up with the large volume of research, medical records, and clinical trials. Watson scales vital knowledge and helps oncologists. By combining attributes from the patient's file with clinical expertise, external research, and data, Watson for Oncology identifies potential treatment plans. [It then] ranks identified treatment options and provides links to supporting evidence for each option to help oncologists as they consider treatment options for their patient. Watson for Oncology draws from an impressive corpus of information, including MSK curated literature and rationales, as well as over 290 medical journals, over 200 textbooks, and 12 million pages of text.

Also using AI to improve medical diagnosis are Enlitic in San Francisco. Here deep learning techniques are being honed to deal with the age-old problem of 'turning images, lab tests, patient histories, and so forth into a diagnosis and proposed intervention'. So great is the potential that Enlitic Founder and CEO Jeremy Howard believes that the application of AI machine learning presents the 'biggest opportunity for positive impact' that he has seen in more than 20 years in the medical field.

Enlitic aggregates all of the information from a patient, including CT and X-ray scans and their medical history, and converts it into a mathematical representation that can be added to and compared with data from other patients. Early results strongly suggest that the system will not only prove faster in reaching a diagnosis, but will achieve a higher level of accuracy than human doctors.

For example, in one batch of scan data used in an Enlitic test, about 7 per cent of the patients given a clean bill of health by a human radiologist were later found to have cancer. And yet, when the company's deep-learning algorithm performed a diagnosis on the same batch of scans, there was not one mistake. A transition to AI diagnostics should therefore result in cancers being detected earlier, so improving patient outcomes and saving healthcare costs.

Already most routine blood tests are analyzed by a machine. Not that many years from now, the development of AI diagnostic systems means that many more medical tests are likely to follow suit. Indeed, by 2030 or maybe even 2025, most patients are likely to favour the opinion of an AI consultant over a traditional doctor who interprets their scan data solely with human eyes and a human brain.

While the potential to use AI systems to improve medical diagnosis is very considerable, there will inevitably be barriers to overcome. Not least, in addition to advancing the requisite technologies, the crowdsourcing of patient data is likely to prove a significant challenge. In the past few decades, healthcare may have gone digital, with most medical cameras and scanners linked to computer-based recording systems. Even so, healthcare currently discards more digital information than any other industrial sector, with at least 90 per cent of the data captured in most hospitals being erased within a week of its capture. In part this is because storing video files has traditionally been very expensive. But it is also because there are currently strict limitations on the pooling of patient information into potentially meaningful Big Data sets.

Right now, only Sweden has the necessary legal structures to facilitate the mass crowdsourcing of healthcare data in a manner that will allow the full benefits of AI medical diagnosis to be reaped. Fortunately, this will not prevent companies like Enlitic, IBM and their partners from making significant progress in data-driven medicine. Yet their work will inevitably be limited until we collectively decide to take a more holistic view.

Highlighting the challenge, in 2011 the McKinsey Global Institute predicted that the US healthcare sector alone could save about $300 billion every year 'by using Big Data creatively and effectively to drive efficiency and quality'. Such savings would amount to around 8 per cent of national healthcare expenditures, and were estimated to be possible even before taking into account future developments in crowdsourced AI medical diagnosis.

Today many people share their lives on social media, and their search histories with Google, but not their anonymized medical data with the broader human community. Though given the extraordinary health benefits on offer, I do suspect that many countries will soon adopt the same approach as Sweden. Not least this may be the case when high-speed, low-cost genomic sequencing becomes very widely available – a subject to which we will return in chapter 9 on postgenomic medicine.

THE ARTIFICIAL EYE

A great many facets of AI involve the development and application of artificial neural networks. These operate not by following programmatic rules, but by learning to interpret and classify input patterns in order to generate the appropriate response. In the human brain, billions of cells called neurons exchange tiny electrical impulses to establish the patterns of connection that enable us to feel, think and remember. Artificial neural networks function in a similar fashion, and are our current best attempt to mirror biological cognitive activity.

The first neural network was called the Perceptron. This was built at Cornell University in 1960 by AI pioneer Frank Rosenblatt, and was taught to recognize foot-high letters placed before an array of photocells. Today, the Perceptron's neural net descendants are able to recognize the far more complex patterns that lie at the heart of many of the AI developments already discussed in this chapter. Increasingly, such patterns include images of everyday objects and human beings, with AI vision recognition advancing very rapidly indeed.

The key thing to appreciate about vision recognition systems is that they turn the previously unstructured data that exists in all still images and videos into a potentially valuable source of information. Today, most mainstream vision recognition systems are limited to activities like tracking cars by reading their license plates, or extracting web links from 'quick response' or 'QR' codes. Any company or individual wanting to keep track of an object and its status therefore has to label it with a barcode or QR sticker, or else attach a radio frequency identification (RFID) tag that can be read by a special scanner. But in the future, AI systems will be able to recognise all of the objects in any picture and correlate that information with all other verified sightings. When coupled with appropriate networks of CCTV cameras, tomorrow's AI technologies will subsequently be able to track most objects on the planet.

In the United Kingdom alone there are already about 5 million active, public-facing CCTV cameras. In the United States, the figure is more like 30 million. Whether you view these figures as scary or reassuring, it is undeniable that we have already created the camera networks that will enable future vision recognition AIs to keep a very close eye on things.

Soon AI vision recognition systems will be intelligently monitoring most production lines by processing the feed from all cameras in the plant. Retailers will similarly be able to keep track of every item in their store by using an AI to monitor their camera feeds. Such developments will explode the popular myth that the Internet of Things (IoT) relies on providing objects with their own Internet connection. In the near future, to 'get online' most objects will simply need to be recognizable and recognized in a sequence of image captures. In turn, this will have significant implications for the operation of most manufacturing, logistics and retail organizations. More widely, it will also impact the packaging and marketing sectors, as creating products that are easy for vision recognition AIs to identify will become a high priority.

One of the major applications of AI vision recognition will be in augmented reality (AR). AR systems, including the Hololens headset from Microsoft or the experimental Google Glass visor, take digital information and mix it with our vision of the real world. Around 2025, AR systems with AI vision recognition capabilities are likely to be used by at least some shoppers to identify every item in view. This could mean that, when you visit a store, the AI in your AR glasses or contact lenses will be able to overlay information on your vision to indicate which goods are cheaper in the shop next door.

Alternatively, a diminished reality AI app will be able to visually erase the worst value products in order to save you the mental strain and distraction of seeing them in the first place. Already diminished reality systems have been created that can remove unwanted objects from video feeds. Just as future virtual assistants will save us from the horror of all of our e-mails, so future vision recognition AIs will help to limit the number of things on which we need to focus our eyes and brains.

AI FACE RECOGNITION

Future vision recognition systems will not be limited to the identification of inanimate objects, with accurate face recognition set to explode as an AI application in the early 2020s. Just one pioneer in this area is a software developer called Cognitec, which describes itself as the 'face recognition company'.

Cognitec already sell facial recognition systems that are used for a wide range of applications that include fraud detection, photo indexing and login authentication. For example, when it comes to detecting fraud, the company's FaceVACS-DBScan software allows government agencies and other bodies 'to find duplicate faces in multi-million photo databases within seconds'.

Hardly surprisingly, FaceVACS-DBScan is already being employed by those who fight crime. As Cognitec explain, 'law enforcement professionals can identify individuals in crime scene photos, videos, stills and sketches by matching facial images against the agency's mugshot repository'. There is even a mobile FaceVACS-DBScan client for Android devices that allows agents to take suspect photos at the crime scene and compare them against central databases.

Cognitec also sell a product called FaceVACS-VideoScan, which can track and identify faces in live video streams in real-time. This clearly has a wide range of security and commercial applications, some of which go beyond the recognition of singular individuals. As the Cognitec website outlines:

> The application detects people's faces in live video streams or video footage and stores video sequences of cropped faces (face streams) for each appearance in front of a camera. Anonymous analysis of all face streams over time allows the software to compute people count, demographical information, people

movement in time and space, and to detect frequent visitors and crowds. FaceVACS-VideoScan also performs real-time identity checks against image databases to find known persons and alert appropriate staff.

A second major pioneer of face recognition is Facebook AI Research (FAIR), which has the mission of 'advancing the field of machine intelligence and developing technologies that give people better ways to communicate'. As Facebook's chief technology officer, Mike Schroepfer, commented in June 2015, 'one of the core problems of the modern age is just how much information is out there. We suffer from this, we can't actually pay attention to everything because we don't have the time'. In the future, Schroepfer hopes that Facebook AI systems will be able to automate tasks like looking through hundreds of baby photos to find the best one.

Some of FAIR's vision recognition research has already resulted in a final released product. For example, an app called Moments can scan a set of images, group them by occasion, identify Facebook friends within the images, and share them directly with those people. In order to achieve this, Moments includes vision recognition algorithms that can match faces across different photos to know who was at a particular event.

For many years, Facebook has been very keen to get users to tag the people who appear in a photo, with the result being a database of tens of billions of pictures featuring one or more named individuals. Facebook claims that its AI technology can already interrogate its entire image database to cross-match a single face in less than five seconds. It also asserts that its technology can recognize faces with 98 per cent accuracy (which compares to the 98.75 per cent accuracy claimed by Cognitec).

To train its neural networking algorithms, Facebook uses a database of photographs of famous people that it calls *Labeled Faces in the Wild*. This is a collection of 13,000 images of celebrities photographed with different hairstyles, different outfits, sometimes wearing spectacles, and other variations. *Labeled Faces in the Wild* is so significant a resource that it has been shared with several other AI vision recognition research teams as a machine-learning tool.

Increasingly, visual biometrics are going to complement usernames and passwords in confirming our identity to machines. Signalling this direction of travel, Windows 10 already includes a biometric authentication facility called 'Windows Hello' that lets people unlock their computer with a facial recognition scan.

In time, AI vision recognition may even become accurate enough to be used to verify financial transactions. In fact, in July 2013, a company called Uniqul in Finland announced that it had created the 'first facial recognition payment system'. In theory, this will allow items to be paid for simply by staring at a single camera. Meanwhile in China, in June 2015 Tsinghua University and a technology company called Tzekwan announced that they had created the first vision recognition ATMs. These still require a user to insert a bank card to withdraw cash, but replace the entry of a PIN code with a facial scan.

At some point in the future, reliable, mass-market, face-only payment systems may be developed. These would probably rely on multiple face recognition verifications, with each target image sourced from a different camera. Future retailer AIs could, for example, continuously monitor a number of CCTV cameras stationed both inside and outside the building in order to verify the identity of each customer when they approach the store, as they enter, while they are browsing, and at the till. Each of these single face recognition attempts may be only about 99 per cent accurate. However, even using existing technology, if a person can be consistently identified by four independent cameras, the aggregate chance of a recognition error falls to about 1 in 41 million. It could therefore be deemed safe to base a payment system on multi-camera vision recognition technology. Although, given that future bioprinters will be able to reprint human faces, vision recognition payment systems may never get off the ground!

CROSSING THE FINAL FRONTIER

Another form of narrow AI that will transform our lives is language translation. Already Google Translate does a pretty effective job of turning words in one language into 90 alternative tongues, with anybody now able to visit translate.google.com to get some text or a web page freely translated. The Google Translate smartphone app even translates spoken language, as well as the text within an image using its WordLens feature. The latter allows a user to hold up their smartphone in order to see road signs, packaging, menus and other text elements translated into another language.

Google Translate already translates over 100 billion words a day. The results are still not perfect, but for most purposes are perfectly serviceable, and on some occasions they are very good indeed. I have, in fact, spoken to some human translators who now start an assignment by applying Google Translate, before tidying things up using their own brain. We should also not lose sight of just how quickly Google Translate – as well as similar apps like iTranslate, Waygo and iHandy – are continuing to improve.

To cite just one example of an evolutionary milestone, in June 2015 a new update was added to Google Translate that improved its handling of casual communications. The update was developed by crowdsourcing the input of the 100,000+ volunteers who serve as members of Google's Translate

Community. Via pooling the linguistic knowledge of these individuals, Google Translate learnt how to improve its translation of many common phrases that fail when translated word-for-word.

Also developing a language translation AI are Huawei in China, with the company's 'Noah's Ark' innovation facility intent on creating a next-generation 'universal translator'. As lead researcher Professor Yang Qiang told the BBC in October 2014, his team's goal is to create software that translates 'by meaning, rather than by syntax'. As the Professor further explained, this should result in the English joke 'Why did the chicken cross the road? To get to the other side' being translated into the Chinese joke 'How do you put an elephant into a refrigerator? You open the door and then you put it in'.

Perhaps not surprisingly, Microsoft are also working on language translation technology, with a preview of Skype Translator released just before this book went to press. The app translates Skype audio or video calls in near-real-time, with an on-screen transcript of the conversation displayed in addition to a speech-to-speech audio translation.

In aggregate, the work of Google, Huawei and Microsoft will fairly soon result in AI systems that can effectively bridge the final logistical divide that separates human communities. In the past 20 years, the Internet may have digitally removed the barriers of time and distance, with people worldwide now free to exchange information anytime and anywhere. Yet language still places a stark obstacle between most people on the planet and most others. If future AI systems deliver no more than the removal of this linguistic buffer, they will therefore have achieved a great deal. I am also certain that this will happen, with most voice- or text-based electronic communication systems likely to offer direct access to a very effective language translation function by 2025 at the latest.

AUTONOMOUS VEHICLES

In addition to delivering virtual assistants, language translation, and a whole host of vision recognition technologies, in the 2020s AI will facilitate the mainstream rollout of autonomous vehicles. In anticipation of this, four US states – Nevada, Florida, California, and Michigan – have now passed laws that permit autonomous vehicles to drive on their roads. The UK Government has put in place similar arrangements, as well as announcing a review of highway legislation. In theory, this will update the UK Highway Code and Ministry of Transport (MOT) vehicle test criteria to account for driverless cars by the Summer of 2017.

Autonomous vehicle technology is likely to arrive in phases. Already some cars feature automated systems that include adaptive cruise control and self-parking. By 2020, such transportation hardware will be joined by semi-autonomous vehicles capable of taking over from a human in certain situations, and which will allow commuters to safely catch up on their e-mail or attend to their make-up while crawling along in traffic.

Around 2025, many experts expect 'high autonomy' vehicles to start becoming available. These will be able to operate autonomously for large portions of a journey, but will require the constant presence of a human driver able to rapidly take control when required. This means that high autonomy vehicles will still feature a steering wheel and other controls, which in turn makes it pretty certain that they will look very similar to the cars of today.

By 2030, motoring is set to more fundamentally change, with driverless vehicles arriving on the market that no longer feature human controls. Such fully autonomous vehicles will be able to transfer people from A to B with no human intervention, and will even be able to undertake road trips with no passengers onboard. This means that, in 15 years time, it will be possible to use your smartwatch to summon your car to come and pick you up, or to send it on a mission to collect those items you left fabricating at a local digital manufacturing bureau. Well OK, a friendly humanoid robot will probably have to help out loading your latest 3D printouts into the trunk. But we will get to such matters in the next chapter.

Many companies have developed autonomous vehicle technologies and prototypes, including Google, Baidu, Tesla Motors, Audi and Daimler. To date, Google's self-driving vehicles are probably the most famous, and have already self-driven over 1 million miles on public roads. Google's fleet currently includes both modified Lexus SUVs, as well as prototype vehicles designed from scratch to be self-driving. The latter have sleek bodies with integrated sensors, as well as removable human controls as a safety feature for test 'drivers'. The first prototype Google driverless car was completed in December 2014.

Taking a more iterative approach, Chinese search engine rival Baidu is working with automobile manufacturers to develop a range of automation technologies to be included in traditional vehicles. Tesla Motors is similarly incorporating driverless systems into its Model S, with Elon Musk reporting in 2014 that his company would be selling 'fully autonomous vehicles' within 'five or six years'. Working incrementally toward this goal, in October 2015 Tesla released version 7.0 of Tesla Autopilot. This allows a Model S to 'steer within a lane, change lanes with the simple tap of a turn signal, and manage speed by using active, traffic-aware cruise control'. As Tesla further explain:

> Tesla Autopilot relieves drivers of the most tedious and potentially dangerous aspects of road travel. We're building Autopilot to give you more confidence behind the wheel, increase your safety on the road, and make highway driving more enjoyable.

While truly driverless cars are still a few years away, Tesla Autopilot functions like the systems that airplane pilots use when conditions are clear.

Also adding autonomous vehicle technology into an existing product is Audi, which has achieved great success with its A7 'piloted driving' concept car. In January 2015, this drove itself 550 miles from San Francisco to Las Vegas to star at CES 2015. Not to be outdone, in May 2015 Daimler Trucks provided the first demonstration of its Freightliner Inspiration Truck – the first commercial vehicle to be licensed for testing on public highways. Make no mistake, developmental autonomous vehicles are now off the drawing board and driving down our roads.

The reasons for developing autonomous vehicles are multitudinous. For some, the vision is to create cars that will allow their owners to focus on driving only when they want to, with their time freed to concentrate on other things when in traffic or cruising down a highway. For other companies, such as Google, the creation of driverless vehicles is more about engendering radical change. As the search giant explains:

> Our goal is to transform mobility by making it easier, safer and more enjoyable to get around. We believe that the full potential of self-driving technology will only be delivered when a vehicle can drive itself from place to place at the push of a button, without any human intervention. For example, there are many people who are unable to drive at all who could greatly benefit from fully self-driving cars.

With sufficient AI behind the virtual wheel, fully autonomous vehicles will have the potential to save fuel by maximizing driving efficiency, and to improve road safety by reducing accidents. In 20 years time, people may indeed wonder why it was ever thought safe to release over a billion tonnes of moving metal out into the world under human control.

Due to bad weather, mechanical failure, the actions of pedestrians, and those remaining human drivers, there are bound to be occasions when even autonomous vehicles will have accidents. Though when this happens, driverless cars are likely to collide more efficiently than those under human control. This is because a lot of good decisions will be able to be made in the second or two of processing time available after the inevitability of an impact has been calculated, but before the actual crash.

For example, future online autonomous vehicles should be able to rapidly and accurately predict the likely human injuries and vehicle damage that will result from a choice of alternative impacts, determine the location and cost of relevant healthcare and vehicle repair facilities, and swerve as appropriate. Two driverless cars about to collide may hence agree on the best angle of impact. Whether you or another person are injured in a future road accident may subsequently depend on a split-second AI assessment of which of you has the rarest blood type or the best health insurance, as well as who will sustain the injuries most easily treated by the surgeons currently on duty in nearby hospitals.

When they have an accident, autonomous vehicles should also be able to make their own insurance claims. More broadly, autonomous vehicles are extremely likely to purchase their own insurance, and on a daily or hourly basis depending on their location and other constantly changing risk factors. The days of selling annual motoring policies via TV advertising really have to be numbered.

Today, every insurance company I have ever spoken to about driverless cars has its attention focused on the thorny issue of who will be liable when autonomous vehicles are involved in an accident. Without doubt, this will require substantial thought. But insurance companies should also be asking why a machine smart enough to drive itself will require a human being to purchase insurance on its behalf, let alone on a fixed, annual basis. In other words, we and they need to recognize that as the AI present in vehicles becomes more advanced, so they will become 'autonomous' in a great many ways. Future cars are extremely likely to take on activities beyond the scope of 'mere' driving. We subsequently need to appreciate that 'driverless cars' will rapidly evolve into one-tonne, four-wheeled robots with highly sophisticated AI brains.

Something else we need to develop an awareness of is the potential for autonomous vehicles to be hacked. In the future, a car thief may be able to stay at home and release a virus that will infect target vehicles and cause them to drive to his or her location. Terrorists may also attempt to hack autonomous vehicles in order to cause mass accidents.

In theory, security will be good enough to prevent these things from happening. Yet given that all commercial software currently deals with bugs and security risks via a policy of release-and-patch, we had better be prepared for carjacking going wireless. In fact, in July 2015, Chrysler became the first automobile company to announce a major recall – of 1.4 million Jeeps – when a software vulnerability was discovered that allowed hackers to take control of the steering, transmission and brakes by patching into the onboard media system.

TOWARD AGI

So far this chapter has focused on narrow AI developments that can automate administrative mental tasks, provide a new computing and customer interface, mine Big Data, diagnose illness, track objects and people, remove barriers to human communication, and drive a vehicle. Over the next few decades, all of these areas of narrow AI application are likely to very rapidly advance, and as this happens we should also expect them to converge and merge. In turn, this ought to allow broader forms of general artificial intelligence (AGI) to evolve.

It would, after all, be far more helpful if we all had just one AI in our life rather than several, with a single virtual assistant not just operating our PC and smartphone, but also able to drive our vehicle, monitor our health, and run our smart home. Given that the age of isolated computing is well and truly dead, we really should not expect future AIs to operate in today's narrow application niches.

Several research teams worldwide are now focused on the creation of AGIs, and it would be staggering if, in the next 20 years, their efforts are not pooled with those currently focused on narrow AI. Progress in areas such as language processing and vision recognition is, after all, going to prove essential for the creation of future AGIs. As illustrated in figure 4.1, we may therefore chart a network of research and development activities that will drive future AGI development, and which will one day all be plugged together.

As figure 4.1 highlights, the developments that will drive the evolution of AGI are already very wide ranging. Clearly those research teams developing specific things like Big Data analytics and driverless car technologies are contributing to the metaphorical primeval soup out of which future AGIs will evolve. But so too are those companies who are developing next generation microprocessors. As we saw in chapter 2, developments in synthetic biology will additionally open up opportunities to create and grow new forms of organic computing hardware that may provide radically new platforms for AGI development. We really should not assume that all future forms of AGI will depend on a silicon semiconductor infrastructure. Human intelligence relies on living, organic

ARTIFICIAL INTELLIGENCE

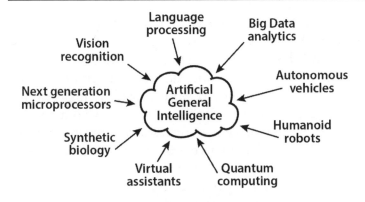

Figure 4.1: The AGI Evolutionary Ecosystem.

'wetware', and it may well turn out that such hardware provides the best or only method for constructing an artificial consciousness.

QUANTUM HORIZONS

In addition to the emergence of biological circuitry, developments in quantum computing could also have major AI implications. Whereas conventional computers use miniature transistors to store and process binary digits known as 'bits', quantum computers store and process information in 'qubits'. Qubits are represented by quantum states, such as the spin direction of an electron or the polarization of a photon. They also have the peculiar ability to exist in more than one state – or 'superposition' – at exactly the same point in time.

Such a quantum mechanical effect is particularly weird. But what it means in practice is that, by attaching a probability to each superposition, a single qubit can be used to process or store a value of '1', '0', or any of the infinite number of decimal values in between. This should make future quantum computers orders of magnitude more powerful than conventional digital hardware, and exceptionally good at parallel processing. In turn, this may make quantum hardware the most suitable for the creation of those neural networks on which AGI development currently depends.

Many companies are taking an active interest in quantum computing. These include Microsoft – which runs a research lab in the area called Station Q – and IBM, who have been investing in quantum computing research for many years. In April 2015, IBM announced that it had created the first computer chip featuring a qubit array. This is made from metals that become superconducting at very low temperatures, and like most quantum hardware has to be operated at just a fraction of a degree above absolute zero.

Also doing pioneering work in quantum computing are D-Wave Systems in Canada. The company first demonstrated a 16 qubit quantum machine in 2007, and in 2011 sold a $10 million, 128 qubit quantum computer called the D-Wave One to a Lockheed Martin research facility. In May 2013, it was announced that a 512 qubit D-Wave Two had been supplied to NASA, Google and the Universities Space Research Association (USRA) to establish the Quantum Artificial Intelligence Lab (QuAIL) at the Ames Research Center in California.

At present QuAIL is focused on demonstrating the future potential of quantum computing to solve currently unfathomable aeronautic optimization problems. Yet in doing so, it may also lay the groundwork for far broader quantum AI developments. In June 2015, D-Wave announced that it had created a 1,000 qubit processor that will form the heart of the D-Wave Three. As the press release explained:

> At 1000 qubits, the new processor considers 2^{1000} possibilities simultaneously, a search space which dwarfs the 2^{512} possibilities available to the 512-qubit D-Wave Two. In fact, the new search space contains far more possibilities than there are particles in the observable universe.

THE SINGULARITY LOOP

The last few pages may have caused you to question just what future AGIs will be capable of, and unfortunately that is an impossible question to answer. It is like asking what an unknown future human being will be able to achieve, and nobody can sensibly answer that. All that we can reasonably predict is that AGIs will one day be capable of undertaking any mental activity, and that will include activities way beyond human comprehension.

The need to undertake activities beyond human comprehension is in fact one of the main reasons for the development of AI in the first place. Stephen Hawking, Elon Musk and others may worry that future AI will get out of human control. But my own concern is that we will not develop AI fast enough in order to make developments like local digital manufacturing a reality. If ten or more billion humans are to have a hope of a prosperous future, then we have to achieve the currently impossible – and to arrive at the Singularity – before this century is out. And getting there demands the creation of AIs with beyond human intelligence.

Fortunately, the more we improve AI, the more it will further evolve. This is because the creation of more sophisticated AIs will permit the development of more sophisticated technologies (such as improved conventional microprocessors, biological computers and quantum chips), so in turn allowing even more sophisticated AIs to be created. As illustrated in figure 4.2, a positive feedback system termed the 'Singularity Loop' may therefore prove to be our salvation. As the figure highlights, AI really is unique as a new technology as its development may catalyze not just its own evolution, but also that of most other areas of technological innovation.

In addition to helping make possible those nine other Next Big things explored in this book, future AIs may one day even provide us with an improved understanding of our-

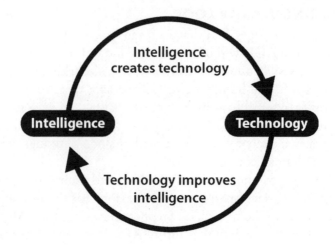

Figure 4.2: The Singularity Loop.

selves. This is because it may always prove impossible to truly comprehend the biological machine of the human brain-and-body combination via self-examination. Almost certainly, only by creating a new race of intelligent, synthetic citizens will we be able to obtain an external perspective on the human animal and the extraordinary civilization we have created.

Given that our species is driven by an evolutionary need to survive, as well as an intense curiosity to pursue the unknown, it seems highly likely that we will relentlessly develop more and more advanced forms of AI. There may be many hurdles and dangers to overcome, but by-and-large we are rather good at overcoming obstacles. Once we get reasonably competent at creating sophisticated AIs, there will also be the opportunity to transplant them into android bodies so that they can assist us with far more mundane tasks than saving ourselves and the planet. In the next chapter, we will therefore anticipate the rise of humanoid robots.

5
HUMANOID ROBOTS

December 2015 saw the much-anticipated release of the seventh *Star Wars* film *The Force Awakens*. After more than 30 years, this blockbuster brought many old favourites back to the big screen, including Luke Skywalker, Han Solo and Princess Leia. The latest episode of the space saga also reunited us with C-3PO, a golden, humanoid 'protocol droid' programmed to fluidly interact with its human masters.

When the first *Star Wars* movie was released in 1977, nobody had built a bipedal robot like C-3PO that could walk, talk and observe social protocols. Yet by the time that *The Force Awakens* hit movie theatres, developmental humanoid robots intended to possess these capabilities did exist. By the 2030s, such mechanical servants are also destined to be a common sight in many companies, and should be available for sale or rental to the general public.

The construction of humanoid robots will present scientists and engineers in many disciplines with a myriad of technical and conceptual challenges. This chapter will therefore report on the work of some of those pioneers who are already labouring to make humanoid robots a reality. However, before we turn to such practicalities, it is worth stepping back to ask just why a large percentage of future robots are likely to be humanoid. Many people's favourite

Star Wars droid is the stumpy, two-or-three-wheeled cylinder known as R2-D2. So why are a significant proportion of our future synthetic citizens more likely to resemble C-3PO?

THE OTHER 50 PER CENT OF JOBS

There are four, key reasons to create humanoid robots, with the first being a logical consequence of ongoing industrial automation. Looking back across the 20th century, we can chart the rise of mass production, with factories tooled-up to produce identical goods in great quantities. Initially this was achieved using dumb machines that performed a single function. But then, with the arrival of computing, opportunities started to exist to create smarter machine tools that could be programmed and reprogrammed to undertake a wider and wider range of tasks. In 1961, the 'Unimate' became the first ever such industrial robot to go into service when it started working on a General Motors production line.

According to estimates from the International Federation of Robotics, around 225,000 industrial robots were sold in 2014, which represented an increase in shipments of 27 per cent compared to the previous year. It also means that, by the end of 2014, the world had an operational stock of about 1.46 million industrial robots, with this figure expected to reach 1.95 million by the end of 2017. In 2013 (the last year for which figures are available), global industrial robot sales were worth about $9.5 billion, with the entire robot-related marketplace (also including software, peripherals and systems engineering) valued at around $29 billion.

As these statistics indicate, robotics is already a very big business. Industrial robots are now also relied upon to undertake a fairly wide range of activities. These include moving things around within factories and warehouses, assisting doctors with surgery, and helping to manufacture vehicles, electrical goods and many other consumer appliances. In fact, in automobile manufacturers like General Motors, pretty much all of the tasks that may possibly be undertaken by a robot *are now undertaken by a robot*. This means that around half of all 'labour' in a modern car plant is already robotic.

To achieve higher levels of physical automation, companies in the future will have to employ robots to do jobs that machines cannot currently undertake. Such jobs invariably require high levels of independent mobility and dexterity. They also demand the ability to work in environments designed for human occupation, and with tools and other artifacts fashioned exclusively for human use. The first reason to develop humanoid robots is therefore to reduce business costs by automating activities that are impossible for a non-human entity to perform.

EMOTIONAL ENGAGEMENT

In addition to being able to labour in the realm of *Homo sapiens*, humanoid robots have more potential than any other mechanical apparatus to physically interact with people on an (apparently) emotional level. The second reason to create humanoid robots is therefore to narrow the communications divide between human beings and machines.

No factory worker seriously pauses to share a moment with a robot co-worker that welds car parts. Yet, as several experiments have already demonstrated, when a humanoid robot turns its head toward you, let alone holds out a hand, a shared context for communication is established. A great deal of human communication is non-verbal, and hence only readily replicated via a humanoid form. If people are to engage successfully with robots in service industries including healthcare and old-age care, we will therefore need to develop hardware that has arms and a head, and which in many cases also features a bipedal locomotive arrangement.

The fact that humanoid robots are more endearing to people than other mechanical forms was first noted in 1970 by a Japanese robotics professor named Masahiro Mori. In the article in which he made this assertion, the Professor also observed that there was a limit to how humanlike a robot ought to become. Specifically, Mori proposed that an 'uncanny valley' existed in robot design, with most people's response to a humanoid robot abruptly shifting from empathy to revulsion as its physicality approached, but failed to attain, a lifelike appearance.

Several studies have since proved Mori's proposition, with most people finding robots that look like animated corpses to be disturbing and creepy. While we should anticipate many future robots to be humanoid, we should therefore not expect them to look like flesh-and-blood people. Just what future humanoid robots will look like when they are bio-printed or otherwise self-assembled out of living cells will therefore be interesting to discover.

TOWARD ARTIFICIAL CONSCIOUSNESS

The third reason to create humanoid robots is to advance artificial intelligence. As already discussed, from a human perspective it will be useful to develop humanoid robots that are capable of living in our world and interacting with human artifacts. The other way around, humanoid embodiment may also be necessary to facilitate the evolution of AIs that can actually understand the human realm. In fact, humanoid embodiment could even be a prerequisite for the development of artificial emotions or artificial consciousness.

While AI systems like IBM's Watson are very impressive, truly intelligent entities without an independent, physical form do not yet exist. And they may never exist. Quite possibly, a cognitive machine without a body will never be able to understand the physical world, as it will be unable to interact with it first hand and to learn from that experience. A prerequisite for the creation of an artificial general intelligence may therefore be for it to be embodied in a robot.

To benefit from the experience of embodiment, a future AI's body need not necessarily be humanoid. However, if we want our future synthetic citizens to empathize with our species and to share human values, we will probably find that providing them with a humanoid form is the most sensible and potentially the safest route forward.

The fabrication of truly intelligent entities within artificial humanoid bodies raises all kinds of questions. For example, if a machine is to develop true intelligence and consciousness, will it need to experience pain and a sense of its own 'mortality'? The 2015 film *Ex Machina* even asked if sexuality may be a prerequisite for the development of an artificial consciousness. As Alex Garland, *Ex Machina*'s writer and director, argued in an interview:

> Embodiment – having a body – seems to be imperative to consciousness, and we don't have an example of something that has a consciousness that doesn't also have a sexual component. If you have created a consciousness you would want it to have the capacity for pleasurable relationships, so it doesn't seem unreasonable that a machine have a sexual component. We wouldn't demand it be removed from a human, so why a machine?

As I argued in the last chapter, the ultimate reason to develop synthetic beings may be to craft new forms of intelligence that are capable of progressing our civilization toward the Singularity. If this proves to be the case, then such AIs will need to understand the world and ourselves at least as well as we do. Many future AIs may be able to teach us a thing or two due to their unique ability to view the world through entirely non-human eyes. And yet, I suspect that the wisest forms of AI will require a great affinity with pre-Singularity humanity it they are to help us on the journey toward a positive future. In turn, this will probably necessitate the wisest forms of AI to occupy a human form at least some of the time.

MEDICAL SYMBIOSIS

The first two drivers of humanoid robot development derive from the requirement to interface with humanity and its legacy infrastructure, while the third may constitute a critical phase in the evolution of AI. In contrast, the fourth and final reason to create humanoid robots is to symbiotically develop robot technology in parallel with future human and transhuman medicine.

Even with the best human brains and AIs tasked with the job, it will take a great deal of time and effort to develop robust and reliable humanoid robots capable of working day-in and day-out in an industrial or consumer setting. Our current human hardware evolved from lower forms of life over millions of years, and yet we are potentially tasked with mirroring the process in well under a century.

While humanoid robots are being created in future research labs, doctors and medical technologists in other facilities will be labouring to improve the health and mobility of the injured, the disabled and the elderly. As we shall see in chapters 9 and 10, some medics and AIs will even be working to proactively evolve humans into enhanced 'transhumans'. If roboticists focus on developing humanoid creations, very significant possibilities will therefore exist for a cross-fertilization of ideas, technologies and components with those at the forefront of medical science and engineering. A good set of legs developed for a robot may, for example, also prove useful in assisting a disabled person to walk.

The above idea may sound quite far-fetched. It is therefore worth noting that, in November 2015, Honda began leasing its first 'Walking Assist Device' to hospitals and other medical establishments. This revolutionary robotic technology attaches to a human patient as a rehabilitation aid. It also arose out of research into artificial bipedal motion undertaken to help create Honda's ASIMO humanoid robot that we will look at in a few pages time. The first productive cross-over between humanoid robots and human healthcare has therefore already taken place.

As bioprinting, synthetic biology and nanotechnology converge into sophisticated forms of local digital manufacturing, so it is likely that we will start to fabricate robots – humanoid and otherwise – out of biological or once-biological materials. In parallel, we are also likely to start fabricating replacement human tissues and even entire body parts using exactly the same methods. It therefore seems implausible that (humanoid) robots and future medical technologies and practices will develop in splendid isolation.

RESISTANCE OR CONSUMER DEMAND?

I can imagine that some readers may be finding this chapter more than a little disturbing. Even setting aside the possibility of wearing an identical set of legs or eyes to your neighbour's android, what I have so far explained is the very real future possibility for humanoid robots to replace human workers. The last chapter raised the similar possibility of disembodied AIs undertaking activities previously reserved for the damp, grey matter of a human brain. You may therefore be asking why we will ever choose a future in which such things are likely to occur.

Since the Luddite textile workers protested against the introduction of mechanical looms in the early 19th century, people have feared for their livelihoods in the face of new technology. There can also be no doubt that, across history, many jobs have disappeared following the invention of new types of machine. And yet, in general, most new technologies have had a positive economic and even social impact, and have led to more prosperous lives for the majority.

Of all the future technologies that could displace human workers, humanoid robots may appear to be the most threatening. Humanoid robots will, after all, be constructed specifically with the intent of undertaking activities that cannot be automated via any other means. Even so, I suspect that, 20 or 30 years hence, humanoid robots will be welcomed far more than they will be resisted.

Ask most people if they would want to 'replaced' by a robot, and they will say 'no'. Yet ask the same person if they would like to live in their own home in their old age, and the majority will choose this option over end-of-life residence in a nursing establishment. Given our ageing population, the cost of providing in-home care for all old people is likely to prove prohibitive. But selling or renting humanoid robot care workers is likely to be cost-effective just a few decades from now. Indeed, in Korea and Japan, it is already starting to be recognized that robots – humanoid or otherwise – will play a major role in old age care in the 2020s and beyond.

Alongside microfabricators, humanoid robots are additionally likely to serve at the heart of the local digital manufacturing revolution. Future developments in synthetic biology, nanotechnology and 3D printing will one day make it possible to manufacture both raw materials and final components in pretty much any location. Yet such materials and components will usually need forming or assembling into final goods, while replacement parts for damaged items will need to be fitted. Sometimes such construction and mending will be undertaken by a growing clan of human 'makers'. But I suspect that humanoid robots will also be involved as a kind of general purpose 'technological glue' that will link and make practical many other new technologies and manufacturing developments.

In addition to providing the cheap 'labour' that will prove critical in the re-localization of both mass manufacturing and a return to mass product repair, humanoid robots may also help to save resources in other ways. In the last chapter I mentioned the development of autonomous vehicles as the future of personal transport. Given the endeavours of Google, Baidu, Tesla, Audi, Daimler and others, I am certain that we will see bespoke driverless cars on the roads in the 2020s and 2030s. Yet I am equally convinced that many future car journeys undertaken without a human driver will have a humanoid robot at the wheel rather than an embedded AI. A key payback of creating a mass market for humanoid robots could well be that automobiles and many other common machines will be able to remain dumb. In turn, this will remove any 'necessity' to scrap a very great deal of perfectly functional conventional technology.

I suspect that many people will also prefer to own traditional vehicles and domestic appliances that a humanoid robot can sometimes operate, rather than buying-in to a world of smart objects that offer no option for human control. Our choice in the future may therefore be to welcome humanoid robots into our world, rather than rebuilding our world for occupation and operation by a whole host of other AI-enabled technologies.

HONDA'S GRAND VISION

Back in chapter 3 I noted how the future may be shaped by forging grand visions called 'attractors'. At a national level, the most powerful attractor ever realized was John F. Kennedy's pronouncement to land an American on the Moon. However, when it comes to the business world, one of the most powerful attractors so far actively pursued was established by Honda in 1986. Here the future vision was to develop a 'two-legged humanoid robot' that could 'genuinely help people' and function in human society. At the time most scientists and engineers considered this to be an impossible dream. Nevertheless, Honda had the considerable foresight to recognize the future potential of humanoid robots, and as a result has become a world leader in the field.

As illustrated in figure 5.1, Honda has created a humanoid robot called ASIMO that can walk on many different kinds of surface, run at up to 9 kilometres an hour, hop forward on one leg, turn smoothly, and climb stairs. ASIMO can also comprehend and respond to voice commands, and has the ability to grasp and manipulate objects with its hands. The latter feature tactile and force-feedback sensors, as well as the independent control of each digit. Due to these sophisticated features, ASIMO is able to communicate using sign language, and can even unscrew the top from a bottle and pour the contents into a cup.

Honda began its crusade by building mechanisms with just an abdomen and two legs. Back in 1986, the first of these was labelled the E0. This was capable of walking, although it took 5 seconds to make each step and could only proceed in a straight line. Between 1987 and 1993, six further iterations were created (the E1 through to the E6), the last three of which helped Honda to develop stable bipedal walking and the ability for a robot to climb stairs. Starting in 1993, Honda next developed three fully humanoid robots that ranged from the P1 – which stood an imposing 1.88 metres tall and weighed 185 kilograms – through to its P3 model at 1.57 metres and 130 kilograms. In the year 2000, Honda then introduced ASIMO, or its 'Advanced Step in Innovative Mobility', whose current incarnation is 1.3 metres tall and weighs 48 kilograms.

In 2011, ASIMO transitioned from being an 'automatic machine' to an 'autonomous machine'. This means that ASIMO now has 'the decision-making capability to determine its behavior in concert with its surroundings'. Specific AI developments that allowed ASIMO to become autonomous included a 'high-level postural balancing capability' (which means that ASIMO can quickly put out a leg

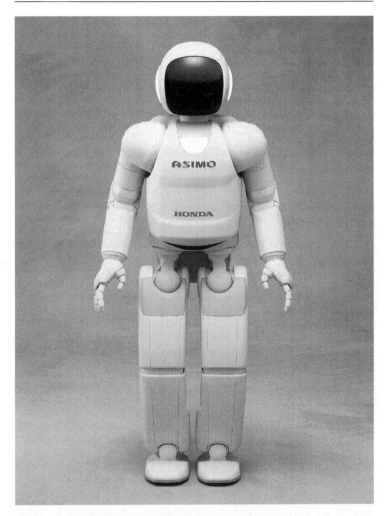

Figure 5.1: Honda's ASIMO. Image courtesy of Honda.

to stop itself falling over); an 'external recognition capability' (that allows ASIMO to gauge and predict the movement of people around it); and a capacity to 'generate autonomous behavior' without operator intervention. In aggregate, these capabilities allow ASIMO to change its behavior to accommodate the intentions of others. For example, ASIMO will step aside to allow a human being to pass, or may alter its trajectory to walk around them. ASIMO now also has integrated face and voice recognition, so allowing it to know who is giving it spoken commands.

To see how remarkable ASMIO has become, just watch the videos available via world.honda.com/ASIMO/. Certainly the robot remains a research-project-in-progress, with a great deal of work required on the AI side before it can be sold or rented for practical applications at a reasonable price. Honda's engineers have, however, already demonstrated the expertise, patience and determination necessary to one day attain their vision of building a viable humanoid robot that can 'coexist with people'.

NAO, PEPPER & ROMEO

Also at the forefront of humanoid robot development is Aldebaran in France, a company with a belief 'in a future where interactive robots will accompany humans in their work, leisure and family life'. Aldebaran was founded in 2005 by Bruno Maisonnier, who as a child had dreamt of building humanoid robots as a 'new species for the benefit of humankind'. The formation of Aldebaran was hence the means to achieve Maisonnier's childhood vision, although the company was bought by the Japanese telecommunications giant SoftBank Mobile in 2012.

In 2006 Aldebaran produced its first robot. Called NAO, this two-foot humanoid was intended to become a 'daily companion' in the home that would learn how to 'help people best'. But to date, NAO has been restricted to educational and research applications, with the robot having become a common piece of hardware used in schools to teach programming, and in universities as a robotics research platform.

In 2008, NAO became the standard player to feature in the annual RoboCup Soccer League competition. The hardware next received a major technological upgrade to 'NAO Next Gen' in December 2011, and is still expected to become available for home use in the future. By the end of 2015, over 7,000 NAO robots had been sold into the education and research community, with the price starting at around $8,000.

In June 2014, Aldebaran and its SoftBank Mobile parent unveiled a second robot called Pepper as 'the world's first personal robot that reads emotions'. Pepper attempts to achieve this ambitious feat by using cameras and other sensors to gauge the facial expressions and body language of those around it. The intention is that Pepper can serve as a companion, as well as providing front-of-house service in shops or other businesses.

Pepper glides around on a skirt-like pedestal, above which it has a humanoid torso, head, arms and hands. The 1.2 metre tall, 28 kilogram robot is already greeting customers in some Japanese retail outlets, and went on limited sale to the general public in June 2015. So great was the interest that the first 1,000 Pepper robots were all snapped up in under a minute. This rapid sell-out repeated when further batches of 1,000 robots went on sale in July, August and September 2015. The price on each occasion was a $1,600 upfront payment plus $200 monthly data and insurance fees. In October 2015, an enterprise version of Pepper – called 'Pepper for Biz' – also became available for rental for $440 a month. If this all sounds very reasonable, it should be noted that SoftBank wants to make Pepper affordable. It hence plans to sell or rent the robot at a loss for four years, with major revenues expected in 20 or 30 years time.

In common with Honda, SoftBank is clearly investing in humanoid robot development as a long-term proposition. To this end, just before the first Pepper robots went on sale it founded a company called SoftBank Robotic Holdings. This has 20 per cent of its shares held by Chinese e-commerce giant Alibaba, and a further 20 per cent owned by the Taiwanese Foxconn (who manufacture for Apple). There can be no doubt that big business is starting to take the future of humanoid robots very seriously indeed.

Since 2009, Aldebaran has also been working on a prototype research robot called Romeo. This 1.4 metre fully-humanoid hardware was created for the Romeo Project, which was initiated and funded by the French Government's Department of Competitiveness, Industry and Services (DGCIS), the Government of the Ile-de-France region, and the City of Paris.

The goals of the Romeo Project included the creation of 'an interactive, open and modular mechatronic and software platform' for the development of a 'personal assistant robot'. By 2012, this and other initial goals were deemed to have been met, and the Romeo 2 Project was launched in November 2012. This second phase has a total of 16 partners – including both commercial companies and universities – and has four years of public funding to develop the Romeo platform for 'personal assistance applications'.

If you are wondering just what Romeo may be able to accomplish, the project's website provides a helpful scenario. This envisages a Romeo robot occupying an apartment with a man who lives alone and who has moderate mobility and memory problems. During the day Romeo keeps track of the man's medication, prepares food for him, warns of a potential danger when the man leaves a cooking pot unattended, and summons assistance when his master does not wake from a nap. A remote operator then takes control of Romeo to give the man a shake, which thankfully wakes him up.

ATLAS & R5

Another very robust humanoid robot has been created by Boston Dynamics – a company purchased by Google in December 2013. Their particular humanoid creation is called Atlas, and is described on its web page as:

> ... a high mobility, humanoid robot designed to negotiate outdoor, rough terrain. Atlas can walk bipedally leaving the upper limbs free to lift, carry, and manipulate the environment. In extremely challenging terrain, Atlas is strong and coordinated enough to climb using hands and feet [and] to pick its way through congested spaces.

Atlas is a very imposing, industrial humanoid robot that is unlikely to evolve into a home-help suitable for putting a child to bed and reading them a good night story. However, when it comes to working in hostile, real-world environments, the robot's 1.8 metre, 180 kilogram frame very much comes into its own.

For example, several 'copies' of Atlas competed and faired very favourably in the DARPA Robotics Challenge (DRC). This was launched in 2011 following the nuclear disaster in Fukushima, Japan, and required participants to 'develop robots capable of assisting humans in responding to natural and man-made disasters'. The finals were held in June 2015, with robot contenders required to drive vehicles, walk through rubble, trip circuit breakers, turn valves and climb stairs.

Of the 23 teams that took part in the DRC finals, seven fielded an Atlas robot. Of these, a team from the Florida Institute for Human and Machine Cognition (IHMC) came second. However, the top prize of $2 million went to a robot called DRC-HUBO entered by the Korean Advanced Institute of Science and Technology (KAIST). Like Atlas, DRC-HUBO was designed for functionality and robustness over aesthetics. It even has wheels on its knees that allow it to crouch down and drive forward in order to increase stability when completing certain tasks.

Another humanoid robot was notable by its absence from the June 2015 DRC finals, although it did participate in earlier heats. As shown in figure 5.2, this was created by

NASA, is known as either 'R5' or Valkyrie, and is the successor to a previous NASA humanoid robot called 'R2'.

R5 is distinguished from many other humanoid robots by its soft coverings (allowing it not to feel cold to human touch), and stands 1.9 metres tall. The hardware was designed by NASA's Johnson Space Center (JSC) both to complete disaster-relief manoeuvres in the DRC competition, as well as to assist with future deep space exploration. For example, prior to a manned mission to Mars, a contingent of R5 robots could be sent to the red planet to prepare a base.

In June 2015, NASA announced that it was seeking teams from the robotics community to further develop R5. The specific intention is to loan R5 units to two DRC teams for two years, along with a $250,000 research grant and NASA technical support. As we shall see in the second half of this book, sophisticated, autonomous humanoid robots are likely to be a prerequisite for future space business and space exploration. It is therefore good to learn that NASA is seeking the best possible help in its quest to make such robots a reality.

GOING OPEN SOURCE

For decades if not centuries, wiley inventors have striven to build humanoid machines. In the 1950s, 1960s and 1970s, those with serious intent attempted to construct mechanisms dedicated to walking and even climbing stairs. But unfortunately for these enthusiastic pioneers, human beings do not walk or climb by engaging a dedicated mechanism. Rather, most variants of human locomotion are a process of 'controlled falling over', with the body thrown off-balance and caught by an extended limb in a repetitive cycle.

A robot capable of 'controlled falling over' needs an array of sensors that can constantly measure the orientation and motion of its body parts. It also needs to be able to read data from these sensors in real-time, calculate feedback loops, and constantly adjust its servos in a manner that will maintain

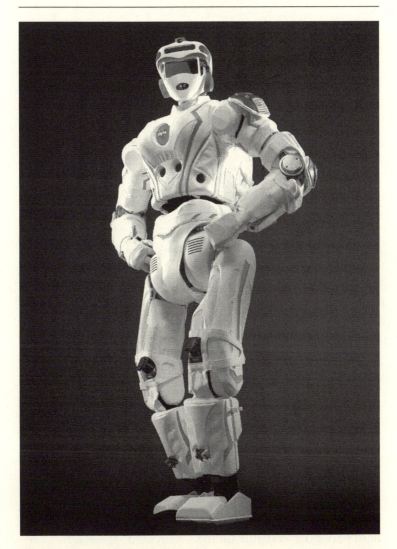

Figure 5.2: NASA's Valkyrie (R5) Robot. Image courtesy of NASA.

and disturb its balance as required. It should therefore come as no surprise that the very significant progress made in humanoid robotics in the past decade or so has been associated with the development of low-cost, high-capacity, lightweight computing power, together with the emergence of sophisticated, low-cost motion sensors (that were often developed for smartphones). Advancements in vision and voice recognition AI – as discussed in the last chapter – have also assisted significantly.

Once a mechanical body has been created with sufficient servo motors, sensors and processing power, the focus in humanoid robot development largely shifts from the hardware to the software realm. ASIMO, NAO, Pepper, Romeo, Atlas, DRC-HUBO and R5 are all testament to this fact, with the research emphasis in all cases now resting on improving the programming and AI. This is also very exciting, as it means that humanoid robot development has the potential to go open source and to be crowdsourced out in the wilds of the Internet. Loads of people with great programming skills have time on their hands, and many of them are likely to be prepared to invest their energies in humanoid robot development. The creation of future synthetic citizens is, after all, an exciting cause in which to collectively invest.

The popular uptake of Aldebaran's NAO as an educational and research tool already demonstrates the considerable potential. However, the price of NAO, let alone one of the other developmental robots detailed in this chapter, clearly lies beyond the financial reach of most ordinary citizens. But fear not – projects already exist to address this obstacle, with designs for 3D printable, open source humanoid robot hardware now available online.

The first such initiative was InMoov, which describes itself as 'the first life size humanoid robot you can 3D print and animate'. The InMoov project was started in January 2012 by a French model maker and sculptor called Gael Langevin, and was 'conceived as a development platform for universities, laboratories, hobbyists, but first of all for Makers'. The

HUMANOID ROBOTS

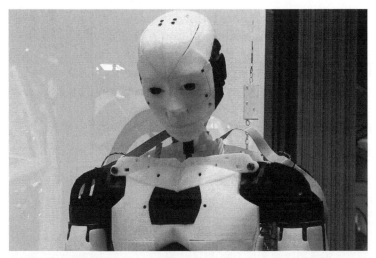

Figure 5.3: The InMoov 3D Printed Open Source Robot.

InMoov robot features a life-size humanoid head, arms and torso, and is illustrated in figure 5.3.

While InMoov's electronics and servo motors do need to be purchased in a conventional manner, all of the robot's plastic parts can be freely-downloaded from inmoov.fr. They can then be fabricated on any personal 3D printer with a 120 x 120 x 120 millimetre build area (or in other words, on any 3D printer costing from about $300 upwards).

A complete InMoov, including all electronics and servos, can be constructed for about $800 – and according to Gael Langevin, putting it together is 'about as difficult as assembling a cupboard from IKEA'.

Already InMoov has been embraced for a variety of purposes, including a project called Inmoov Robots for Good. This is building a remote-controlled version of InMoov that is intended to be operated by children in hospital in order to allow them to take a virtual visit to London Zoo.

A second, 3D printed open source creation is Poppy Humanoid, courtesy of the Poppy Project. This describes itself as 'an open-source platform for the creation, use and sharing of interactive 3D printed robots' and has gathered 'an interdisciplinary community of beginners and experts, scientists, educators, developers and artists'.

Poppy Humanoid stands 84 centimetres tall, and with all of its 3D printed parts, servos and electronics included, weighs 3.5 kilograms. The aim of the Poppy Platform is to develop 'open-source tools for [the] rapid prototyping and flexible experimentation of robotic creatures'. Poppy tools are therefore designed 'to be modular, easy to use, and easy to integrate – providing a set of building blocks that can be easily assembled and reconfigured'.

RESHAPING THE ECONOMY

While all of the humanoid robots detailed on the previous pages are amazing, they are clearly only the very early ancestors of tomorrow's smart, autonomous mechanoids. Things are, nevertheless, moving rather quickly – and not just in terms of hardware and AI development.

In April 2011 I finished writing a book called *25 Things You Need to Know About the Future*. This included a chapter on robots, but featured far fewer real humanoid robots than I have been able to report on just four years later. As has been demonstrated, very large companies – including SoftBank Mobile, Alibaba, Foxconn and Google – have now started to invest in humanoid robots. Just as importantly, journalists, the business community, and the broader business press, have also begun to recognize the looming implications of (humanoid) robots as a Next Big Thing.

For example, in February 2015, the Boston Consulting Group (BCG) issued a report in which it noted how 'the use of advanced industrial robots is nearing the point of takeoff'. Indeed, the BCG report predicted that by 2025:

> . . . the adoption of advanced robots will boost productivity by up to 30 percent in many industries and lower total labor costs by 18 percent or more in countries such as South Korea, China, the US, Japan, and Germany.

The BCG went on to attribute current and future improvements in robot performance to 'advances in vision sensors, gripping systems and information technology'. As BCG Partner Michael Zinser subsequently explained:

> For many manufacturers, the biggest reasons for not replacing workers with robots have been pure economics and technical limitations. But the price and performance of automation are improving rapidly. Within five to ten years, the business case for robots in most industries will be compelling, even for many small and midsized manufacturers.

The BCG predict a global spend on robots of $67 billion by 2025. It has to be noted that this money will be spent on robots in general, and largely not on the humanoid variety. Though given that increased automation will require robots to take on tasks that demand the use of human tools in human environments, it is virtually inevitable that the BCG's research points toward a fairly widespread use of humanoid machines sometime in the late 2020s and beyond.

In the above context, it is perhaps not surprising that author and *Forbes* staffer John Tamny believes that 'as robots increasingly adopt human qualities, including those that allow them to replace actual human labor, economists are starting to worry'. Tamney does, however, also contend that such fears are 'unfounded', as 'abundant job creation is always and everywhere the happy result of technological advances that tautologically lead to job destruction'. Indeed, according to Tamny, robots will turn out to be 'the biggest job creators in history'. Why? Well simply because:

> ... aggressive automation will free us up to do new work by virtue of it erasing toil that was once essential.... With their evolution as labor inputs, robots bring the promise of new forms of work that will have us marveling at labor we wasted in the past, and that will make past job destroyers like wind, water, the cotton gin, the car, the Internet and the computer seem small by comparison. All the previously mentioned advances made lots of work redundant, but far from forcing us into breadlines, the destruction of certain forms of work occurred alongside the creation of totally new ways to earn a living. Robots promise a beautiful multiple of the same.

While I personally would agree with Tamny, not everybody is of the same opinion. And even if robots do positively impact the economy as job creators, nobody disputes that they will lead to the emergence of new kinds of job in the wake of the destruction of previously core forms of human employment. This means that, at the very least, we face a very significant period of economic and social upheaval as we transition to an economy part-staffed by autonomous, smart machines.

In his best-selling book *Rise of the Robots*, Martin Ford cautions that the 'human economy' has always demanded 'routine work', but that in the future such work will 'not be done by humans'. Renowned MIT professors Erik Brynjolfsson and Andrew McAfee agree, and recently explained in *Business Insider* how there is no guarantee that new replacement forms of human labour will be created as rapidly as routine human jobs disappear. There is hence the risk that a generation or two may end up being 'skipped' during a very painful period of transition. The 2030s and beyond are really not going to be a good time for those with few or no qualifications, or who expect to be paid to render an autonomous bulk of unthinking human muscle.

SHARING THE PLANET

Since 1984, the Search for Extraterrestrial Intelligence (SETI) has been using radio telescopes to look for evidence of intelligent life on other planets. Over the years, SETI activities have also attracted much media interest, and on occasion have really captured the public imagination. Given that the project may one day introduce us to an extraterrestrial intelligent species, interest in SETI's work is quite understandable. And yet, in comparison, I continue to be surprised at how little attention we collectively pay to the creation of non-human forms of intelligence here on Planet Earth.

As the last two chapters have hopefully demonstrated, within decades there are going to be new species of highly intelligent 'synthetic citizens' living with us on our first planet. In his excellent, free-to-read book *The Second Intelligent Species*. Marshall Brain – the guru founder of *How Stuff Works* – paints a particularly stark picture of just what this will mean. Whether tomorrow's synthetic citizens are disembodied AIs or robots (humanoid or otherwise), as Brain notes they will take many of the jobs at which millions if not billions of human beings currently labour. You may therefore not be surprised to learn that the subtitle of Brain's aforementioned book is *How Humans Will Become as Irrelevant as Cockroaches*.

In a separate but equally powerful online essay entitled *Robotic Nation*, Brain predicts that robots like C-3PO are a logical future extension of current business automation, and 'will come into our lives much more quickly than we imagine'. His precise prediction is that 'the first machines that [can] see, hear, move and manipulate objects at a level roughly equivalent to human beings' will arrive by 2025, and humanoid robots as a 'commodity' for replacing human staff will be with us around 2030.

Like Brain, I too expect humanoid robots to start walking, working and playing among us from around 2030, with the average model costing the equivalent of a family automobile today. As has already been noted, the short- and long-term economic implications will be very significant. But so too will be the deeper, psychological impact. Humanity has been used to being the most intelligent species on the planet for a very long time indeed. Making the psychological adjustment to being one of the *two* most intelligent species is therefore not going to be easy.

In addition to their economic, social and psychological impact, future humanoid robots and AIs are going to raise a lot of practical issues. For a start, will the owner of a future humanoid robot be held legally responsible for its actions? And if not, just how will future humanoid robots be held to account and insured? Many science fiction authors – and most famously Isaac Asimov – have proposed that future robots will need to be regulated by hard-coded internal 'laws' that will prevent them from harming human beings. Such fictional rules may well become a reality. This said, I suspect that most future disputes involving robots will involve mundane, domestic matters like whether or not it was your neighbour's robot that woke your baby or made a footprint in your lawn.

More fundamentally, we will need to debate whether highly autonomous and even sentient AIs and robots ought to have legal and moral rights. Should, for example human 'masters' always have the option to turn on and turn off future synthetic citizens as they please? Or to access their memories? Or to reprogram them at will? Clearly the current paradigm for the creation of AIs and humanoid robots is to construct them as helpful slaves. But when we begin to make very intelligent synthetic citizens that appear to be sentient – whether they are ever actually gauged to be sentient or not – so the 'slavery model' may have to change.

The above is likely to become a signature debate when we start to craft humanoid robots not just from plastic, silicon and metal, but out of living cells that are fabricated into robotic forms using bioprinting, synthetic biology and self-assembly nanotech. Such fabrication mechanisms are likely to be used in a pretty much identical fashion for both human medicine and robot construction and repair. Will it therefore remain grievous bodily harm when somebody stabs the leg or other organ of a human being, and yet little or no crime if a human owner sticks a knife into exactly the same body part of 'their' humanoid robot?

We really do need to start thinking about this kind of thing and fast. Over the past few decades we may have fairly comfortably adjusted to the mass use of dumb computing devices. But within 20 years, today's dumb computers will have been replaced with cohabiting synthetic citizens. In the future, we will therefore not just own smart technology. It will live with us. And that is something that we will need to learn to live with.

PART III

RESOURCES FROM SPACE

6

SPACE-BASED SOLAR POWER

I have always been fascinated by the Second Law of Thermodynamics. This basically states that every physical thing will decay, with no closed system able to last forever. Put a rabbit in a hutch, and even if you leave it with a supply of water and lettuce, it will sadly expire unless you periodically open the cage and provide further provisions. Those on submarines are similarly reliant on regular trips to the surface to replenish their finite stock of air and other supplies. Closed systems simply cannot go on indefinitely – and this is true whether they are tiny laboratory experiments or entire civilizations.

Over the past four decades, a great deal of research has attempted to focus mainstream attention on the resource constraints of the closed system called Planet Earth. Perhaps most notably, in the early 1970s a project undertaken at MIT for the Club of Rome built a computer model to calculate humanity's current and predicted use of natural resources. The findings were first published in the 1972 book *The Limits to Growth*, with further updates released in 1992 and 2004. The final two books benefitted from more data and increasingly sophisticated computer models. But all three publications carried the consistent warning that the 'ecological footprint' of human civilization now exceeds the 'carrying capacity' of Planet Earth.

Figure 6.1 draws from *The Limits to Growth* data to chart humanity's physical resource requirements against the available supplies of our current, finite planet. As you can see, we started to require more than a single Earth could supply around 1980, and are currently headed toward needing 1.5 planets.

If, after looking at figure 6.1, you are thinking 'well, we are still here – so this analysis must be wrong', then you need to remember that all physical systems can exceed their limits for short periods. Most people can, for example, run very fast for a few minutes or even hours, but will need to rest, drink and eat if their life is to continue. A similar physical restriction also applies on a planetary level. Right now the Earth's natural ecosystem is just about coping with what we are taking from it and throwing back. But we cannot go on exceeding our planet's finite carrying capacity indefinitely.

BEYOND SUSTAINABILITY

Most people concerned about the Earth's finite stock of natural resources advocate the pursuit of 'sustainability'. I too believe that we all need to *consume less and value more* in order to respect the planetary ecosystem that keeps us alive. Even so, I also believe that the pursuit of sustainability will be insufficient to meet the basic resource requirements of future generations.

The concept of sustainability was introduced in 1987 by the United Nations World Commission on Environment and Development (WCED), also known as the Brundtland Commission. In their report *Our Common Future*, the WCED defined 'sustainable development' as 'development which meets the needs of current generations without compromising the ability of future generations to meet their own needs'. Or, in simpler words, sustainability was defined as 'living today without compromising tomorrow'. Such a concept is a highly laudable goal. But, as the Second

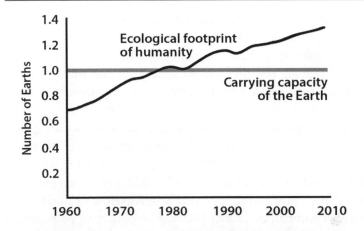

Figure 6.1: Resource Requirements & the Carrying Capacity of the Earth. Based around *The Limits to Growth* data.

Law of Thermodynamics informs us, it is also a physical impossibility.

Current sustainability measures, including recycling and switching to alternative energy sources, will help to preserve our dwindling resource supplies. Yet they can at best only constitute a short-term solution. In addition to consuming less, in the future we will therefore need to go 'beyond sustainability' to find more energy sources and raw material supplies. Or in other words, we will need to open up the closed system of our first planet.

As nanotechnology guru Eric Drexler argues in his book *Radical Abundance*, while *The Limits to Growth* was truly inspirational, its environmental computer model was flawed because it was confined to Planet Earth. It is almost inevitable that sometime this century our civilization will demand an annual quantity of energy and raw materials that our motherworld will be unable to supply, let alone sustain. But this need not be an end game. For as Russian space visionary

Konstantin Tsiolkovsky once noted, 'Earth is the cradle of the mind, but one cannot live in the cradle forever'.

The following three chapters explore future possibilities for obtaining resources from space. As we shall see, the options include harvesting solar energy in orbit, as well as mining both the asteroids and the Moon. None of these things are going to be remotely easy. In fact, they are going to be extremely difficult. Obtaining resources from space will therefore probably only become the Next Big Thing in the second half of this century. This said, there are a growing number of pioneering individuals, research teams and commercial companies already focused on obtaining extraterrestrial resources.

THE SOLAR REVOLUTION

Most forms of energy come from the Sun. For example, plants give us nourishment obtained from sunlight via photosynthesis. All meat also contains solar energy converted by flora somewhere down the food chain. More significantly for industrialized economies, fossil fuels provide power from ancient sunlight stored up by long-deceased plants and animals. Even wind turbines are indirectly powered by solar energy as it heats some parts of the atmosphere more than others and causes the air to move around. Given that our local star is expected to go on burning for several billion years, the development of new ways to extract solar energy is therefore likely to be vital to our future.

Current solar power technologies work in a number of ways. The simplest are solar thermal systems, which use the Sun to heat water for washing or heating. A second set of 'light redirection' technologies alternatively include fiber optic cables or mirrored ducts to distribute the Sun's rays inside a building as a source of illumination.

A third set of 'concentrated solar power' (CSP) technologies employ parabolic reflectors to focus the Sun's heat, which is then used to create steam, drive a turbine, and generate electricity. Alternatively, a mechanical apparatus known as a Sterling Engine can be used to directly produce electricity by heating a gas that moves a piston.

The fourth and final method of producing energy from the Sun is to use photovoltaic (PV) cells to directly turn sunlight into electricity. The involved physics is complex, but basically uses light to trigger a flow of electrons between two layers of semiconducting material.

While each of the aforementioned solar power technologies serve to capture energy from the Sun's rays, they all suffer from a fairly critical suite of limitations. Not least, they do not work at night. Even in the day, their performance is variable and difficult to anticipate due to potential cloud cover and the changing seasons. Large solar power stations also require large areas of land. It is therefore not surprising that the long-term future of solar energy may be in space.

Space-based solar power (SBSP) – also know as space solar power (SSP) – would generate off-world energy by placing solar power satellites in orbit. These future space stations could potentially receive sunlight up to 24 hours a day, and would be unaffected by cloud cover. The sunlight hitting a solar power satellite (SPS) would also not be filtered by the atmosphere or dust contamination, so again increasing the relative generating efficiency of any solar panel receiving it. Taking all of these factors into account, renowned SBSP scientist John Mankins has calculated that 'the available solar energy in space [is] about ten times greater than the best average available at most locations on Earth'.

In his 2014 book *The Case for Space Solar Power*, Mankins goes on to note that the argument for building ground-based solar power stations is often diminished by the 'difference of roughly a factor of three in the solar energy available in summer versus winter'. He also notes that current electrical energy storage systems are less than 50 per cent efficient, so again reducing the viability of ground-based solar power systems that require storage technologies to deliver a constant energy supply.

TOWARD ENERGY FROM SPACE

The idea of placing solar power satellites in orbit was first conceived by an American space engineer called Peter Glaser. As part of a career that included contributions to both the Apollo and Space Shuttle programs, on 25 December 1973 Glaser obtained United States Patent 3,781,467 for his 'method and apparatus for converting solar radiation to electrical power'. This described an arrangement wherein:

> Solar radiation is collected and converted to microwave energy by means maintained in outer space on a satellite system. The microwave energy is then transmitted to earth and converted to electrical power for distribution.

Glaser envisaged solar power satellites with microwave antennas about one square kilometre in area, and which would transmit power to smaller 'rectenna' stations on the ground. As he stressed in his patent, the idea was to meet 'a measurable part of the electrical power requirements of the Earth' by 'a system and method which do not deplete the Earth's natural resources and which are essentially pollution free'.

Glaser's plan included two solar power satellites in equatorial, geostationary orbit 35,786 kilometres above the Earth. At this distance up in space, the satellites would be fixed above a single point on the Earth, and hence would always be able to beam their power to a fixed rectenna on the ground. Each satellite would, however, inevitably sometimes pass through the Earth's shadow, and so the intention was to position them about 21 degrees out of phase, or roughly 12,700 kilometres apart. This would allow both satellites to have a direct line of sight to the same rectenna station, with at least one of them illuminated by the Sun at all times.

Building on Glaser's vision, in the mid-to-late 1970s the US Energy Research and Development Agency (ERDA) – now the Department of Energy (DOE) – spent about $20 million on research into the potential viability of obtaining solar power from space. The work was undertaken in conjunction with NASA, and resulted in a number of high-level designs based around a concept now known as the 1979 SPS Reference System. This would have featured a solar power satellite with a 5 x 10 kilometre array of photovoltaic cells, coupled to a 1 kilometre microwave dish for power transmission. Such a system could potentially have delivered 5 or even 10 gigawatts of power to a rectenna some 10-12 kilometres in diameter.

In 1979 it was estimated that a 300 gigawatt SBSP system consisting of 60 satellite platforms and all supporting infrastructure would have cost up to $3 trillion. It would also have taken around 20 years to construct, and would have required reusable launch vehicles three times bigger than the Space Shuttle to lift its components into orbit. In addition, hundreds of astronauts would have had to labour in space for long periods. Perhaps not surprisingly, the ERDA and NASA proposals for such a system were not accepted. Nevertheless, it is interesting to ponder what the world would be like today if this grand vision had been commissioned and was now in successful operation.

While, in 1980, the US Government rejected any further SBSP investment, broader interest and research was maintained in Japan, France and Canada. The SunSat Energy Council – an organization founded by Peter Glaser and affiliated with the United Nations – also continued to promote the concept. A growing number of conferences, publications and websites additionally kept the idea alive, while the spectre of climate change slowly drove alternative energy sources up the popular and political agenda. Between 1995 and 1997, NASA subsequently decided to look again at SBSP in a study that became known as Fresh Looks.

The Fresh Looks initiative followed a reorganization of NASA in 1994, and had the goal of determining if new technologies developed since 1980 might make solar power from space more feasible. The work was both wide-ranging and comprehensive, and resulted in a 1997 report called *The Fresh Look Study of Space Solar Power*. This included information on more than 30 potential systems and satellites. Sadly, after some initial enthusiasm, NASA again decided not to take things further (although it should be noted that competing interests to potentially develop space nuclear power systems may have influenced the decision).

Outside NASA, the response to Fresh Looks was more positive, and further studies continued to be undertaken. For example, in 2007 the US Department of Defence National Security Space Office presented a detailed analysis of SBSP as a 'potential grand opportunity to address energy security'. Around this time, several SBSP startups were also founded, including the Space Island Group (SIG), Solaren, the Space Energy Group, Planetary Power, Heliosat, Managed Energy Technologies, and the PowerSat Corporation. All of these companies promised to deliver demonstrators and systems for budgets of hundreds of millions or billions of dollars. But unfortunately, none of them got anything off the ground. This sad point noted, we must remember that the financial crisis of 2007-2009 did not assist the pursuit of funding for radically new, extremely expensive, high-risk and long-term undertakings.

SBSP ARCHITECTURES

Between 2008 and 2011, the International Academy of Astronautics (IAA) undertook one of the most significant SBSP initiatives to date. Its final report – *Space Solar Power*

- *The First International Assessment Of Space Solar Power: Opportunities, Issues and Potential Pathways Forward* – incorporated contributions from experts in 10 countries. As the extended title suggests, the IAA study had a wide remit, with its goals being:

> . . . to determine what role space solar power (SSP) might play in meeting the rapidly growing need for abundant and sustainable energy during this century, to assess the technological readiness and risks associated with the SSP concept, and (if appropriate) to frame a notional international roadmap that might lead to the realization of this visionary concept.

One of the major contributions of the IAA report was a detailed analysis of three 'highly promising' solar power satellite (SPS) platforms. The chosen architectures incorporated a wide range of potential options, and according to the IAA were all found to be technically feasible. Indeed, as the IAA noted, 'there are no fundamental technical barriers that would prevent the realization of large-scale SPS platforms during the coming decades'. A brief overview of the IAA's three platform configurations therefore provides a solid foundation for understanding where obtaining solar energy from space may be headed.

As shown in figure 6.2, the IAA's Type I architecture is an updated version of the SPS Reference System developed by the ERDA and NASA in 1979. Here, one or more very large arrays of photovoltaic solar panels are coupled to a microwave power transmission system. The architecture requires a large, stabilized platform that rotates on three axes to keep its solar panels pointed at the Sun, and its power transmission dish targeted at a receiving station on the Earth.

The Type I SPS architecture has the advantage of making use of 'traditional' SPS technologies. It does, however, also

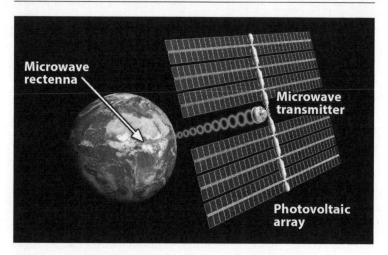

Figure 6.2: IAA SPS Type I Concept.

present a number of practical challenges. These relate to 'the need for an extremely large, high-voltage power management and distribution system on the platform', as well as 'the requirement for substantial up-front infrastructure' both in space and to launch the required hardware.

As illustrated in figure 6.3, the IAA's Type II system transmits energy to the Earth using multiple lasers that target a photovoltaic rectenna. While solar-pumped lasers could potentially be used to directly channel focused sunlight to the Earth, the IAA concluded that electric lasers are likely to prove more effective. The IAA's Type II architecture is therefore based on electric laser power transmission, with sunlight turned into electricity using large but modular photovoltaic arrays. Power is then transmitted to the Earth by a laser incorporated into each module. The lasers would operate in the near-visible part of the spectrum, for example using near-infrared wavelengths.

As shown at the top of figure 6.3, a Type II architecture could comprise a modular but integrated platform, with each

SPACE-BASED SOLAR POWER   171

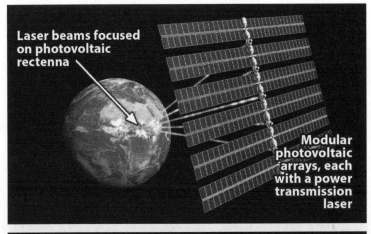

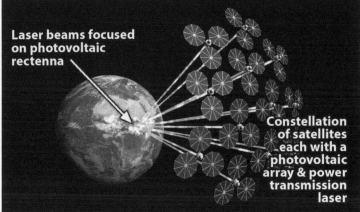

Figure 6.3: IAA SPS Type II Concepts.

station module including its own photovoltaic array and laser installation. Alternatively, as illustrated at the bottom of the figure, a Type II system could be a constellation of small, entirely independent platforms, each with its own photovoltaic array and laser hardware.

Because of its integrated or distributed modular construction, a Type II architecture could become operational in

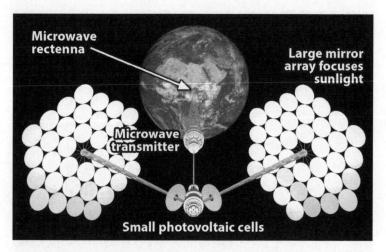

Figure 6.4: IAA SPS Type III Concept.

stages. This means that power could start to be delivered to the Earth as soon as a single solar-array-and-laser module was in orbit, with additional modules added over time. On the other hand, as the IAA argued in their report, 'electric-laser SPS concepts have a significant challenge to compete in terms of end-to-end efficiency with microwave based concepts at power levels greater than approximately 100 MW'. The heat waste from laser as opposed to microwave energy transmission may also be unacceptably high.

The IAA's Type III system is based on a symmetrical optical concentrator or 'sandwich' design that does not require very large photovoltaic panels. Rather, as shown in figure 6.4, vast arrays of mirrors collect and redirect sunlight onto much smaller photovoltaic solar cells. The electricity so created is then transmitted to the Earth via a single microwave link.

The benefit of the Type III approach is that it requires a less complex power management and distribution arrangement than necessary in a Type I or Type II system. This is because captured solar energy is largely moved around via light redirection, rather than through a physical electrical infrastructure. Type III systems may therefore prove the most straight-forward to construct. This said, the concentration of solar energy on the photovoltaic array could create an unacceptably high level of waste heat.

Subsequent to the work of the IAA, in September 2012 an SBPS architecture known as SPS-ALPHA was presented by John Mankin in a report written for the NASA Innovative Advanced Concepts Program. SPS-ALPHA stands for 'Solar Power Satellite by means of Arbitrarily Large Phased Array', and in common with the IAA Type III architecture uses light redirection to focus solar energy onto a photovoltaic array, with power once again beamed to the Earth via microwave. But rather than employing a sandwich dual concentrator design, the SPS-ALPHA features a large number of individually pointed, lightweight thin-film mirrors arranged in a conical form. Such a structure is biologically inspired, and would be created in space from a very large number of modular components that would self-assemble into the single, massive solar power satellite.

In addition to negating the need for a large electrical infrastructure in space, an SPS-ALPHA would significantly reduce the need to launch into orbit or to construct in space very large pieces of hardware. Indeed, according to Mankin, all major components would be both individually small and 'intelligent', with non more massive than 100-300 kilograms. In fact, aside from the propulsion and attitude control modules, each of the seven other types of modular element would weigh 50 kilograms or less, with some as light as 1 kilogram.

The SPS-ALPHA is an ingenious concept that takes the nanotech self-assembly paradigm we looked at in chapter 3 and applies it on a macro scale. Just as individual insects can form into a single colony, so an SPS-ALPHA's 'swarm' of smart, standardized components would self-arrange into a large, functional structure. In turn, this may finally make space solar power an economically viable proposition. Indeed, according to Mankins' calculations, 'it appears that a full-scale SPS-ALPHA, when incorporating selected advances in key component technologies, should be capable of delivering power at a levelized cost of electricity (LCOE) of approximately 9¢/kilowatt-hour'.

GRAND CHALLENGES AHEAD

While the IAA believes that solar power from space is a future possibility, there are clearly a great many technological hurdles to overcome. Or as John Mankins argues in *The Case for Space Solar Power*, solar power satellites are technically possible, 'but the technology is not proven and sitting on the shelf'.

The two key technology areas that need to most significantly advance are wireless power transmission (WPT) and space transportation. In some respects, the first of these has come a long way in the past decade or so, with some smartphones now charged by placing them on a wireless inductive plate. But transmitting energy from geostationary orbit? Well, that is another proposition entirely.

To many people's surprise, experiments in long-distance wireless power transmission have been taking place for decades. Indeed, as long ago as 1964, William C. Brown from the Spencer Laboratory of the Raytheon Company in Burlington, Massachusetts, flew a microwave powered model helicopter. This used a 2.9 metre dish to transmit energy 15 metres to a craft with a 1.8 metre rotor that stayed airborne continuously for 10 hours. As Brown concluded in his explanatory article for the *Journal of Microwave Power*, there appeared at the time to be a wide variety of applications for the technology. And one of these is clearly SBSP.

A more recent pioneer of wireless power transmission is Professor Nobuyuki Kaya of Kobe University in Japan. In 1995 the Professor organized a conference on the subject, and as part of the proceedings used microwaves to transmit about 5 kilowatts of power from a ground transmitter to a rectenna mounted underneath a helium-filled airship.

In September 2008, Professor Kaya also worked with John Mankins and Frank Little of Texas A&M University on a longer-distance experiment. This used 2.4 GHz microwaves to transmit about 20 watts across the 148 kilometres that separate the islands of Maui and Hawaii. Interestingly, the project was sponsored by the Discovery Channel, and in the process garnered much popular interest.

Several other experiments have demonstrated the potential viability of microwave-based wireless power transmission, with researchers in Japan remaining at the forefront of this field. Most recently, in March 2015, scientists from the Japan Aerospace eXploration Agency (JAXA) managed to transmit 1.8 kilowatts of power over 55 metres 'with pinpoint accuracy'.

At present a great deal of the energy sent by microwave transmission is lost before it can arrive at the receiver, with an efficiency of 20 per cent or less not being uncommon. Power loss during wireless energy transmission is therefore a major issue to overcome. This said, in one of his many experiments, pioneer William C. Brown did manage to send 30 kilowatts over one mile with 84 per cent of the power actually arriving.

As noted in the discussion of the IAA's Type II SPS architecture, future solar power satellites may transmit their energy to the ground using laser beams rather than microwaves. Boeing and NASA have recently been working on such 'laser-photovoltaic wireless power transmission' (laser-PV WPT). A private company called LaserMotive is also developing significant expertise in the field, and in 2010 flew an electric quadcopter continuously for nearly 12.5 hours powered wirelessly by laser from the ground. While this was impressive, at present lasers are even less efficient energy carriers than microwaves, with a maximum of 10 per cent of the power surviving transmission.

Even when its technical challenges are overcome, the future transmission of energy wirelessly from space is likely to raise public concerns. After all, what happens if a bird or a plane enters the power beam? Well, according to LaserMotive, with laser-based systems there could be sensors in position that would detect objects in flight and turn off the beam as required.

When it comes to the safety of microwave systems, the power transmission frequencies would have to be those at which the atmosphere is transparent, while the area of the transmitter and receiver are likely to be very large indeed (at many kilometres across). These technical factors mean that the frequency and intensity of the beam at any one location would not cause injury or damage to animals, people or machines. This said, aircraft would surely be heavily discouraged from entering wireless power transmission airspace to prevent any disruption of their electronic systems.

Additional fears may relate to the use of future solar power satellites as weapons due to their ability to 'fire' energy at the Earth. With laser-based satellites, there may perhaps be such dangers. But microwave-based satellites should not pose a threat due once again to the low power of their energy beam at any one location.

OBTAINING ORBITAL ACCESS

Solar power satellites will, quite literally, have to get off the ground. That, or they will need to be constructed in orbit from materials either ferried from the Earth, or potentially mined from the asteroids or the Moon. Whichever alternative may prove the most viable, hardware deployment and maintenance will require very regular access to Earth orbit.

SBSP development will therefore have to go hand-in-hand with major advancements in space transportation.

The provision of space access is currently transitioning from the public to the private sector. Already most satellites are launched by private companies, including Arianespace (which flies its own Ariane and Vega launch vehicles, along with Russian Soyuz hardware), Orbital ATK (formerly Orbital Sciences Corporation, which operates rockets called Antares, Minotaur and Pegasus), and International Launch Services (which flies Russian Proton launch craft).

A relatively new entrant to the launch business is SpaceX, which now provides NASA with a transportation service to the International Space Station (ISS) using its Dragon space capsules and Falcon 9 rockets. Orbital ATK is additionally providing an ISS ferry service using its Cygnus capsules, while JAXA supplies the station using the H-IIB rockets and H-II capsules built by Mitsubishi Heavy Industries (MHI). However, currently only the Russian Space Agency can ferry both humans and cargo to and from the ISS. Here the hardware comes in the form of the tried-and-tested Soyuz rockets and capsules manufactured by OAO S.P. Korolev Rocket and Space Corporation Energia (otherwise known as RSC Energia).

Another major player in space transportation is Boeing, who are working on NASA's new Space Launch System (SLS). This will be the most powerful rocket ever constructed, and will be capable of lifting up to 130 tonnes of cargo into orbit. Boeing is also a partner in the United Launch Alliance. This is a joint venture with Lockheed Martin Space Systems, and currently provides the Atlas and Delta family of rockets that carries US Government payloads into space. The United Launch Alliance is additionally working on a new launch vehicle called Vulcan.

In 2017, both SpaceX and Boeing are due to start flying astronauts to and from the ISS. Working toward this goal,

SpaceX has developed a seven-seater Dragon 2 or 'Crew Dragon' capsule that will fly atop its Falcon 9 rocket. Meanwhile Boeing is working on a seven-seat capsule called the CST-100 that will be launched by a United Launch Alliance Atlas 5.

What the last page or so should tell us is that an increasing handful of commercial companies are developing the orbital access technologies that SBSP will require. It would indeed be very surprising if a future SBSP initiative did not involve at least some of the aforementioned organizations, or else the China National Space Administration (CNSA) and its main contractor, the state-owned China Aerospace Science and Technology Corporation. Many other companies – such as Virgin Galactic and Amazon founder Jeff Bezos' startup Blue Origin – have or are developing space transportation systems. But SBSP will demand orbital access on a grand scale, and suitable technology will take time, expertise and copious amounts of money to develop. I am therefore pretty certain that we know who will be involved in providing the orbital transportation if a SBSP mission launches anytime before 2030.

Looking further ahead, it is possible that SBSP will have to wait for a future generation of space access technology that is currently not even off the drawing board. For example, it may be that SBSP will be contingent on the creation of 'space elevators'. Here, as shown in figure 6.5, a cable would stretch from a ground station up to an orbital space platform in geostationary orbit 35,786 kilometres above the Earth. People and cargo – including solar power satellite components – would then travel to and from space in an elevator carriage.

It is even possible that the power generated by a solar power satellite could be transmitted down to the ground using a space elevator's cable. This could potentially happen if the cable included fiber optic strands capable of containing wireless laser power transmissions, or else superconductors

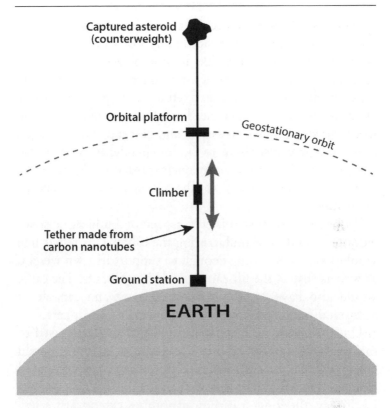

Figure 6.5: Space Elevator Concept.

able to carry high-voltage electricity over 35,786 kilometres without a significant loss of power.

As shown in figure 6.5, a space elevator's cable or 'tether' would extend way beyond the orbital platform to a counterweight that would keep it taut and its centre of gravity motionless relative to the Earth. Launching such a counterweight is likely to prove prohibitively expensive, and so a suitable asteroid would probably be captured and used for this purpose. At the other end of the cable, the most likely position for a ground station would be a high altitude location somewhere on the equator.

Although the term 'space elevator' is in fairly popular use, the system could be more accurately described as a 'space ladder'. This is because, unlike in most elevator systems, the carriage would not be hauled up and down by a moving cable. Rather, with only one, fixed tether in place, the space elevator carriage would need an independent climbing mechanism to allow it to ascend and descend. The carriage's climber would also have to be independently powered – perhaps by solar cells or a tiny nuclear reactor. Alternatively, power may be beamed to the climber using lasers or microwaves.

Probably the most difficult aspect of building a space elevator would be manufacturing the cable. This tether line would have to be strong enough to support its own weight, as well as that of the lift carriage and its contents. The cable would also have to be corrosion resistant and capable of withstanding the extreme cold of the upper atmosphere.

Until recently, no material existed that could be used to make a space elevator cable (a steel wire, for example, would collapse under its own weight). But a cable manufactured from carbon nanotubes is now at least a theoretical possibility. Today, the longest carbon nanotubes extend just a few centimetres. But with new production technologies it may one day be possible to craft cables tens of kilometres long that could be woven into a space elevator tether. Such cables may perhaps be self-assembled downwards from a suitably captured asteroid until they reached the ground.

SBSP PIONEERS

As we have seen, the technical challenges that need to be overcome before a single solar power satellite can enter operation are really quite staggering. Yet so were the challenges faced by NASA when President John F. Kennedy asked them to land an American on the Moon. Building SBSP systems will be extremely difficult, but not impossible. So, which nation will be the first to rise to the challenge?

Well, the smart money is currently on the Japanese. Certainly, over the past five years, several reports have indicated that the Japanese Space eXploration Agency (JAXA) is very actively conducting research into what it terms 'space solar power systems' (SSPS). For example, in 2010, the agency published an interview with Yasuyuki Fukumuro, the scientist then leading the research. As he explained:

> There are many technological challenges to solve before SSPS can be implemented. However, in principle, we are getting close to the stage where it is feasible, and we have just moved from the study phase to the technology demonstration phase. Researchers have started preparation for the world's first demonstration of 1Kw-class wireless power transmission technology, and are aiming for practical use in the 2030s. At this point, you could say that Japan is leading the world in SSPS research.

In April 2014, Susumu Sasaki – an Emeritus Professor at JAXA who spent most of his 41-year career researching SBSP, published an article in *IEEE Spectrum*, the journal of the Institute of Electrical and Electronics Engineers. Within, he revealed that JAXA 'has a technology roadmap' for 'a series of ground and orbital demonstrations', and which could lead to the development of a 1 gigawatt commercial SBSP system sometime in the 2030s.

An option apparently being considered is the construction of geostationary space hardware similar to the IAA's Type III configuration, except with its two parabolic mirrors flying free in space with no physical links between them. The idea is that only one of the reflectors would ever enter the Earth's shadow, hence allowing solar energy to be continuously generated by the photovoltaic array that would fly in formation between the two colossal reflectors.

In his 2014 article, Professor Sasaki envisaged microwave power from space being received by a 3 kilometre artificial island constructed in Tokyo Bay. As he went on to explain:

> A massive net is stretched over the island and studded with 5 billion tiny rectifying antennas, which convert microwave energy into DC electricity. Also on the island is a substation that sends that electricity coursing through a submarine cable to Tokyo, to help keep the factories of the Keihin industrial zone humming and the neon lights of Shibuya shining bright.

According to the roadmap outlined by Professor Sasaki, JAXA is planning a technology demonstration mission by 2018. This would transmit several kilowatts of power from a satellite in low Earth orbit. The agency then envisages a 100 kilowatt test satellite operating by 2021, a 200 megawatt commercial satellite by 2028, and a 1 gigawatt commercial power plant (supplying about the same output as a current nuclear power station) by 2031. Finally, by 2037, the commercial SBSP industry would be bringing a new solar power satellite on stream every year.

Japan's rising interest in solar power from space is being driven by two key factors. Firstly, the country lacks both fossil fuel reserves and enough land to construct large-scale wind or ground-based solar power installations, and is hence dependent on imported energy resources. Secondly, the accident at the Fukushima nuclear power plant in 2011 promoted an exhaustive search for an alternative means of clean energy generation, and SBSP may in future fit the bill.

While at least some people in Japan consider SBSP to be a near future possibility, in the United States the previous interest in the technology appears to have peaked. As already noted, those start-up companies that hoped to progress the concept in the mid-noughties met resistance in the form of the global financial crisis. In more recent years, the United States has also started to exploit very substantial reserves of unconventional fossil fuels, and this too has diverted business and political attention away from SBSP.

Over in Europe, in 2010 Airbus Defence and Space (formerly EADS Astrium) was apparently seeking partners to fly a demonstration solar power satellite. Its intended prototype was deemed capable of transmitting about 10 kilowatts of power using an infrared laser, and could have flown 'within the next five years'. As of November 2015, while the company's website was still reporting the development of 'new systems and technologies for transferring orbital solar energy to Earth', there was no evidence of further progress.

In March 2015, the Xinhua News Agency — China's official press agency — reported that 'Chinese scientists are mulling the construction of a solar power station'. In the article, Wang Xiji from the Chinese Academy of Sciences (CAS) noted that SBSP technology must be acquired before 'fossil fuels can no longer sustain human development'. As he went on to note, 'whoever obtains the technology first could occupy the future energy market. So it's of great strategic significance'.

In a 2010 report, members of the Chinese academies of science and engineering suggested that China was capable of building an experimental space solar power station by 2030, and a commercial facility by 2050. Quite how advanced plans may be toward this potential goal is publicly unknown. Right now, Japan appears to be the leading player in the field. But as Wang Xiji noted, the first nation to deliver a source of long-term, clean energy from space will put itself in a very powerful position.

NEAR-SINGULAR HORIZONS

Space-based solar power is a long-term future vision that will be extremely difficult – if not impossible – to realize with even the best of today's cutting-edge technologies. In turn, this means that SBSP is only likely to become possible when we get very close to the Singularity. Only AI minds may be able to fathom the engineering complexities of obtaining solar power from space. And only future methods of microfabrication may be able to construct the required gigantic hardware out of new nanotech materials that currently do not exist.

Back in Part I of this book we looked at a range of possibilities for future local digital manufacturing (LDM). In those earlier chapters, 'local' was defined relative to the position of actual or potential human customers. But it is equally possible that 'local' may one day be a measure of the orbital position of a solar power satellite.

We will certainly not be able to launch solar power satellites in one piece, or even in very large sections. Construction in space is therefore going to be required, and it may turn out that it is easiest to microfabricate solar power satellites from scratch out of very basic raw materials. Possibly, the necessary building supplies will be ferried up from the Earth in rockets or by space elevators that have also been microfabricated. Though I personally suspect that, in the long-term, most solar power satellites and other space infrastructure will be made in space from materials mined from the asteroids or the Moon. Such extraterrestrial excavation is also the subject of our next two chapters.

Space-based digital manufacturing (SPDM) has, in fact, already started to happen, with a 3D printer called the Zero-G delivered to the ISS in September 2014. This was developed for NASA by a company called Made in Space, and in December 2014 printed out a ratchet wrench from a digital design transmitted from the Earth. Some space hardware already includes 3D printed components, with parts of Airbus Defence and Space satellites now additively manufactured from aluminium (if admittedly currently on the ground).

In January 2014, SpaceX even flew one of its Falcon 9 rockets with a 3D printed main oxidizer valve, while its Crew Dragon capsule will feature eight SuperDraco engine chambers 3D printed out of the superalloy Inconel. As Elon Musk, the founder and CEO of SpaceX has explained, 'through 3D printing, robust and high-performing engine parts can be created at a fraction of the cost and time of traditional manufacturing methods'. We should therefore expect 3D printing to play an increasing role in improving the performance and lowering the cost of space access technologies. In turn, the more items of space hardware that are 3D printed, the greater the opportunities will become to fabricate or re-fabricate such components in space.

Along with 3D printing, space-based digital manufacturing could also capitalize on developments in synthetic biology. While purely organic off-world components are unlikely, as John Mankins' plans for the SPS-ALPHA suggest, solar power satellites may potentially be self-assembled using 'natural' biological processes. The 'cells' that would be used as the building blocks could possibly be 3D printed in orbit. Alternatively, they may be microfabricated in space using some future form of nanotechnology that is able to turn, for example, asteroid fragments into semi-autonomous, self-positioning and self-interlocking components.

In the future, 'colonies' of solar power satellites built and maintained in 'organic' ways may even be categorized as a form of macro synthetic life. In time, the nanotech microfabrication systems that are likely to form part of such 'creatures' may even allow them to 'procreate'.

Without doubt, the computing systems and sensor arrays required by solar power satellites will need to be highly sophisticated. The massive entities that beam energy to their parent planet may hence become the largest ever species of autonomous, embodied artificial intelligence. Future astronauts may therefore travel into space to farm or wrangle solar power satellites – if, that is, such entities require any human involvement in their creation or operation.

Humanity has always survived and thrived by thinking big to achieve the apparently impossible. We are also already making significant progress toward those Singularity-level innovations in AI and digital fabrication that could allow space-based solar power to become a reality. As progress toward this goal accrues, we are additionally likely to broaden our horizons far beyond Earth orbit. In the next two chapters, we will therefore explore the possibilities for obtaining fresh raw materials from beyond our first planet.

7
ASTEROID MINING

The human race is gobbling up physical resources at an alarming velocity. To highlight this fact, in May 2011 the United Nations Environment Programme (UNEP) published a report called *Decoupling Natural Resource Use and Environmental Impacts from Economic Growth*. As this heavyweight document explained, if left unchecked humanity's annual demand for natural resources will rise to around 140 billion tonnes by 2050. This is almost three times our current rate of resource consumption (which in 2010 stood at 50 billion tonnes), and is 'far beyond what is likely to be sustainable'. Unless we make some dramatic changes, within a few decades we will therefore hit 'Peak Everything'.

As outlined in earlier chapters, a transition to local digital manufacturing could lessen our future resource requirement by enabling more material-efficient fabrication, and reducing the production and disposal of unwanted goods. In addition, innovations in synthetic biology could allow new organic raw materials to be cultivated in almost any location, and from little more than sunlight, air and water. Future nanotechnologies may also allow a wide variety of waste – including much of the garbage that currently lies festering in landfills – to be transformed back into pristine raw materials. As has been the case throughout history, breakthroughs in science and engineering are set to provide at least partial solutions to those apparently insurmountable challenges that lie ahead.

Applying new technologies to help eek out the finite resources of our motherworld is undoubtedly a very sensible short- and even medium-term proposition. Nevertheless, it should not constitute our only long-term survival strategy. As I noted at the beginning of the last chapter, the Second Law of Thermodynamics condemns all closed systems to decline and decay. This means that, even if we adopt every single new technology possible, in the very-long-term the material resources of Planet Earth will still not be able to sustain our civilization.

Some futurists believe that, by the second half of this century, the Earth will only be able to support about two billion people. This is clearly terrifying given that there are currently over seven billion humans on the planet, with around nine billion expected by 2050. Within the next few decades, we will therefore need to choose between two alternative futures. The first would involve a painful process of mass depopulation. Alternatively, the second and more palatable option would involve rising to the challenge of obtaining additional physical resources from space.

Hopefully we will soon commit to a substantial expansion of our material supply horizons. If and when we do, we will also need to choose between two potential new resource frontiers that we could realistically start to exploit before 2100. One option is to begin excavating raw materials from the surface of the Moon. Alternatively, as this chapter will investigate, we may choose to venture into space on a grand asteroid mining crusade.

DEBRIS ON A GRAND SCALE

Asteroids – also known as minor planets or planetoids – are primordial chunks of space debris in orbit around the Sun. They are the material left over following the formation of the

ASTEROID MINING

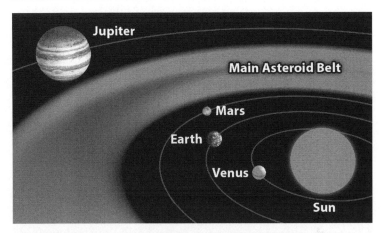

Figure 7.1: The Location of the Main Asteroid Belt.
Celestial bodies not to scale.

planets about 4.6 billion years ago, and range in size from large boulders to hundreds of kilometres in diameter.

As illustrated in figure 7.1, most asteroids are located in the Main Asteroid Belt between the orbits of Mars and Jupiter. Estimates suggest that the Main Belt's inner and outer rims are roughly 330 million and 480 million kilometres from the Sun. This means that, at any one point in time, the closest Main Belt asteroids are at least 180 million kilometres from the Earth, or a good 450 times further away than the Moon.

The Main Asteroid Belt once contained enough dust and rock to form a planet about four times larger than the Earth. But once Jupiter was formed, the gas giant's enormous gravitational pull disrupted the formation of such a planet, so leaving a disparate field of unattached asteroids. In time Jupiter's gravity swept away most of the material in the Main Belt, with the combined mass of free-floating material in this region of space now being less than that of the Moon.

Given that many asteroids are very small indeed, nobody actually knows how many there are in the Main Belt. This said, the count is probably in the tens of millions. According to NASA, we are also pretty certain that the Main Belt contains between 1.1 and 1.9 million asteroids that are over 1 kilometre (0.6 miles) across, with over 200 of these exceeding 100 kilometres (60 miles) in diameter. The largest Main Belt asteroid is Ceres with a diameter of about 940 kilometres (590 miles), although in 2006 it was reclassified as a dwarf planet. Other very large Main Belt asteroids include Vesta and Pallas, with diameters of around 544 and 530 kilometres respectively.

While the Main Belt contains millions of asteroids, it also occupies about 54 trillion trillion cubic kilometres of space. This means that most asteroids in the Main Belt are hundreds of thousands of kilometres apart. The popular image of the Main Asteroid Belt as an impenetrable field of space rock is therefore far from accurate. Rather, the asteroids in the Main Belt are so thinly distributed that numerous unmanned spacecraft have passed through without incident. In fact, if you were standing on one asteroid in the Main Belt, it is very unlikely that you could actually see another unless it was orbiting as a captured moon or formed part of a binary asteroid combination.

As illustrated in figure 7.2, located outside the Main Belt are three groups of near-Earth asteroids (NEAs) called the Atens, the Amors and the Apollos. Technically, an asteroid can be termed an NEA when its orbit brings it within about 195 million kilometres of the Sun, which in turn places its path within about 45 million kilometres of the Earth's orbit.

NEAs are being discovered on an increasingly regular basis, and by February 2015 there were estimated to be least 12,000 such asteroids. Of these, approaching 1,000 are more than 1 kilometre in diameter, with the larger bodies including asteroids named Ganymed, Sisyphus and Hathor.

Many NEAs are potentially hazardous asteroids (PHAs), as their orbits are not just close to that of the Earth, but on

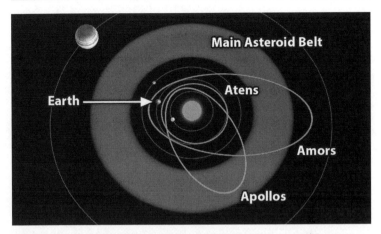

Figure 7.2: The Location of the Near Earth Asteroids.
Celestial bodies not to scale.

occasions actually cross over it. In fact, there are at least 1,400 PHAs that pose a potential risk to the survival of all life on this planet. The future mining of PHAs may hence present an opportunity not just to obtain resource supplies, but to protect ourselves from a cataclysmic collision. The latter may potentially be achieved by mining the PHA to extinction. Alternatively, we could remove enough material to suitably change its orbit, or move the asteroid out of harm's way by detonating explosives, tugging it with a spacecraft, or attaching rocket engines to its surface.

Around 1,500 NEAs are prime candidates for mining operations. In part this is because of their relatively close proximity to the Earth. However, it is also due to the nature of their orbits. Understanding this fact is also quite important, which means that we now need to delve into just a little rocket science.

In space the difficulty in getting from one place to another is determined not just by distance, but by the required velocity change – or 'delta-v' – necessary to shift the mass of a spacecraft or other body between the two locations. The greater the delta-v value, the more propulsion is required, and hence the more difficult the journey is to accomplish.

For example, to escape the Earth's orbit entirely requires a delta-v of 11.2 kilometres a second (km/s), while to get from the Earth's surface to low-Earth orbit (LEO) requires a delta-v of 8.5 km/s. To journey from LEO to the surface of the Moon then requires a delta-v of 6.3 km/s. But to travel from LEO to the easiest-to-access NEA requires a delta-v of just 4.0 km/s, with around 200 NEAs requiring a delta-v of 6.5 km/s or less. The delta-v for returning from many NEAs to LEO is then only around 2 km/s, so making the return of mined deposits a far more attractive proposition than, for example, transporting anything from the surface of the Moon.

What all of the above means is that there are probably between 150 and 200 NEAs that are easier to access than the Moon, and which we could travel to in months rather than years. Any asteroid mining in the remainder of this century is therefore going to take place on an NEA, and not on one of the far more distant and difficult to access asteroids in the Main Belt.

ASTEROID COMPOSITION

In November 2005, a Japanese probe called Hayabusa touched-down on an NEA named Itokawa. In June 2010, it then made a triumphant return to the Earth where it delivered a sample capsule that contained a few small particles of asteroid material. At the time of writing (in November 2015), the precious particles returned by the Hayabusa probe are the only asteroid materials to have ever been directly analyzed by human scientists.

Before we commence asteroid mining, we clearly need to develop a very good understanding of what asteroids are made of and their likely value. Given that, at present, we have only directly sampled one asteroid, you may imagine that this presents quite a challenge. However, via a technique called 'spectrophotometry', a telescope can be used to measure the wavelengths of sunlight that reflect from an asteroid's surface. By comparing these readings with those of known rocks and meteorites here on the Earth, reasonable deductions can then be made concerning the asteroid's surface composition. Having applied this method for many years, astronomers have satisfied themselves that asteroids are made of a variety of materials, with the majority spectrally classified as 'C-type', 'S-type' or 'M-type'.

C-type or 'carbonaceous' asteroids – also known as chondrites – are the most common, and account for about 75 per cent of all asteroids analyzed to date, including about 50 per cent of NEAs. Like most other asteroids, C-types are largely comprised of rock, which in this particular instance is made from dark-coloured clays and silicates. This means that C-type asteroids are rich in carbon, and can also be water bearing. Our knowledge of such asteroids is based not just on telescopic spectrophotometry, but in addition a chemical analysis of some of those 50,000+ 'carbonaceous chondrite' meteorites that have impacted the Earth.

C-Type asteroids are in turn separated into five classes, which are referred to as C1 to C5. Of these, the first are potentially the most valuable, as they contain about 10 per cent water held in a clay mineral matrix, or within hydrated rocks. C1 carbonaceous chondrites also contain between 2 and 5 per cent carbon (in the form of graphite, hydrocarbons and possibly frozen carbon dioxide), together with magnesium salts, sulphur, iron sulphide and the iron oxide magnetite. C2 carbonaceous chondrites may in addition be a rich source of future resources, as they are thought to contain up to 10 per cent pure iron, and up to about 33 and 23 per cent silicon and magnesium oxides respectively.

The second most common category of asteroid is the S-Type, which stands for 'stony' or 'silicaceous'. These chunks of space debris are mainly made of silicates, together with smaller quantities of sulphides, nickel, iron and possibly other metals. In total, about 17 per cent of all asteroids are believed to be S-Type, although they are more common in the near-Earth groups than in the Main Asteroid Belt.

The third main asteroid category is M-Type, standing for 'metallic', although this spectral classification is sometimes also labelled 'X-Type'. This is due to the difficulty in performing telescopic spectrophotometry on the limited light reflected back from the surface of such space rocks, which has called the composition of M-Type asteroids into dispute. Nevertheless, it is still generally accepted that most M-Type asteroids harbour high concentrations of nickel and iron, as well as some silicate deposits (which deliver similar spectral readings). Certainly M-Type asteroids are very dense indeed, with some thought to contain very high concentrations of platinum, as well as deposits of other rare and highly valuable metals. In total, about 10 per cent of asteroids are estimated to be M-Type.

In addition to those materials mentioned above, some asteroids in all three of the most common categories are believed to contain small but significant deposits of other metals including aluminium, cadmium, cobalt, germanium, iridium, magnesium, rhodium, palladium, silver and gold. Many of these are highly valuable, and have a wide range of applications – for example in the manufacture of electronic components.

In addition to C-Type, S-Type and M-Type asteroids, there are also some far rarer categories with entirely different compositions. These include 'V-type' asteroids, which are believed to be made up of basaltic rocks. All V-type asteroids were initially thought to be fragments of the giant Main Belt asteroid Vesta. But then, in 2001, the discovery of V-Type asteroids dissimilar in makeup to Vesta disproved this theory – although the 'V-Type' label remains in use.

Following an orbital visit by NASA's Dawn probe in 2011 and 2012, Vesta was very surprisingly found to have large quantities of hydrogen on its surface. As this unanticipated discovery ought to remind us, while we have already collected a lot of data on asteroid composition, we still have much more to learn. We would not, after all, expect a distant alien species to develop a full understanding of the composition of the Earth simply by analyzing the spectral patterns reflected from its surface.

A final and extremely important thing to appreciate about all asteroids is that their raw materials are likely to be of a very high quality. This is because, unlike Earth-based resources, asteroid deposits have not been subject to a grand-scale planetary metamorphosis and resultant contamination over the past few billion years.

Not least, the gravitational forces of a planet have not had a chance to draw heavy metals down from an asteroid's surface. Metal deposits have also not been oxidized by an atmosphere or living things, while fossilized remains have not had the chance to accumulate. And, of course, asteroids have not yet had their best pickings scavenged by humans. Rather, the material deposits on asteroids have lain undisturbed for around four billion years. Many asteroid deposits are therefore on or close to the surface, with metals in particular expected to be refinery grade. It is subsequently not surprising that asteroids have been termed a paradise for both geologists and prospectors.

THE MARKET POTENTIAL

Raw materials mined from asteroids are going to meet the needs of two very different markets. The first of these will be extraterrestrial, with asteroid deposits – including water and hydrocarbons – initially likely to be used as a source of rocket fuel, and to support the human occupation of space.

As has been noted, launching anything into orbit, let alone releasing it entirely from the grasp of the Earth, requires an enormous amount of energy. Transporting large quantities of oxygen, water or rocket fuel into orbit or beyond is hence likely to remain a complex and prohibitively expensive proposition. The first, great challenge for this century's space pioneers will therefore be to establish an off-world source of these critical substances. This means that water in particular is going to become an extremely valuable space commodity, as it is separable via electrolysis into oxygen and hydrogen, with the former able to sustain humans, and the latter able to be burnt in the presence of oxygen to propel a rocket. We should therefore expect C-Type asteroids rich with water deposits to the first choice for asteroid mining pioneers.

In addition to sustaining space-based activities, first generation asteroid mines are likely to serve the resource requirements of major space-bound infrastructure projects. For example, future space elevators will require vast orbital counterweights that are very likely to be made from asteroid materials (or indeed just one suitably-sized asteroid or cut section thereof).

Space-based solar power stations may also be constructed from asteroid materials. The fact that many NEAs are rich in metals, semiconductors, and other materials required to make electronic components like photovoltaic solar cells, will therefore prove rather handy. Even basic asteroid rock could prove a valuable space-sourced building material, as it could be used to shield human quarters from deadly cosmic radiation.

In addition to solar power satellites, materials mined from asteroids could help to construct and sustain other categories of spacecraft and space stations. Such vehicles and orbital habitats may be used by space tourists, as well as by future scientists and engineers engaged in the development and production of medicines that can only be fabricated in zero gravity. As soon as off-world sources of oxygen and rocket fuel become abundant, the economics of space access are bound to be dramatically transformed. Once this occurs, many more reasons are likely to be found for people to venture into Earth orbit and out across the Solar System.

As well as serving exterrestrial requirements, mined asteroid deposits will in time start to enter the terrestrial commodities market. Initially, some people may be prepared to pay vast sums of money just to own jewelry or other items that include very small quantities of asteroid gold, silver or even plain rock. The first few thousand rings, paperweights or other trophies which incorporate some asteroid materials may indeed become the ultimate gifts or talking points for the super-rich and materially minded.

The first asteroid deposits to enter Earth's commodity markets in larger quantities will be rare and expensive metals including silver and gold, and in particular the platinum group metals (PGMs). The latter comprise platinum itself, together with ruthenium, rhodium, palladium, osmium and iridium. While we may not talk about the PGMs very much, they are used in very small quantities in the manufacture of a large number of industrial products, including electronic components, automotive catalysts, some medical products, jewelry and fuel cells. PGMs are very rare indeed, making platinum more expensive than gold. World production of platinum is currently in the order of only a few hundred tonnes a year.

About 10 per cent of NEAs are rich in the PGMs. Estimates vary, but some platinum-rich, 500 metre wide asteroids have been calculated to contain over 100 times the yearly world output of platinum, and 1.5 times the known world reserves. According to Mining.com, an NEA called 2011 UW-158 is thought contain 90 million tonnes of platinum and other precious metals.

In time, and as off-world operations scale and prices fall, asteroids could provide many of the rarest and most expensive feedstocks required by future microfabricators. This means that, while many future products will be manufactured locally, the rarest and most expensive non-bulk materials used in their fabrication could be sourced from millions of miles away. It is therefore possible that many of today's teenagers will one day own products made at least in part from asteroid materials.

TRILLION DOLLAR ROCKS

If you want proof of the potential scale of the future asteroid mining business, then just pay a quick visit to asterank.com. This is 'a scientific and economic database of over 600,000 asteroids' that uses data from multiple sources (including NASA) to 'estimate the costs and rewards of mining asteroids'. The site features a stunning, animated 3D visualization of its database of near-Earth and Main Belt asteroids, and allows visitors to produce tabular reports by selecting criteria such as 'most cost effective', 'most valuable' and 'upcoming passes'.

According to Asterank calculations, there are 528 asteroids out there (admittedly in the Main Belt) that each harbour in excess of an estimated $100 trillion of natural resources. Meanwhile, a lot closer to home, the 900 metre Apollo NEA '1999 JU3' has a delta-v value of 4.664 kilometres a second, and boasts deposits of nickel, iron, cobalt, water, nitrogen, hydrogen and ammonia worth an estimated $95.02 billion.

Mining 1999 JU3 would be expensive, but the profit from such a venture is estimated at $34.52 billion. If you are interested, the asteroid's next close passes to the Earth will occur in July 2016, December 2020, June 2025 and July 2029. In December 2014 the Japanese Space eXploration Agency (JAXA) launched a probe called Hayabusa 2 to collect samples from this asteroid and to return them to the Earth by 2020. But do not fear. If you miss out on 1999 JU3, an Amor asteroid called 1943 Anteros is estimated to be worth $5.57 trillion with a $1.25 trillion profit – and so is worth a longer trip.

DEEP SPACE LOGISTICS

While the potential gains may be very great indeed, the practicalities of asteroid mining will not be easy. For a start we have to get there, and deep space travel is still very much in its infancy. This said, several small, unmanned probes have now visited asteroids. These include NASA's NEAR (Near Earth Asteroid Rendezvous) Shoemaker, which landed on the S-Type asteroid 433 Eros in February 2001. As mentioned a few pages back, JAXA's Hayabusa landed on an NEA named Itokawa in November 2005, and even returned a sample of it to the Earth.

Other spacecraft that have travelled very close to asteroids, but which have not landed, include the NASA craft Galileo (which got within 1,600 kilometres of 951 Gaspra in 1991 and 2,400 kilometres of 243 Ida in 1993), Deep Space 1 (which flew within about 26 kilometres of NEA 9669 Braille in July 1999), Stardust (which had a 3,300 kilometre encounter with the Annefrank asteroid in 2002), and Dawn (which orbited Vesta in 2011 and Ceres in March 2015). The European Space Agency's Rosetta probe also got within 805 kilometres of the asteroid Steins in 2008, and as close as 3,160 kilometres to the asteroid Lutetia in 2010.

The above missions have proven our ability to rendezvous with and even land on asteroids. Some of the aforementioned craft have also served as testbeds for technologies that may prove critical for future deep space activities. Not least, Hayabusa, Deep Space 1 and Dawn all featured ion engines – a form of electric propulsion that ionizes and accelerates xenon atoms, rather than burning rocket fuel.

Ion engines do not create a great deal of thrust (the Dawn probe took four days to accelerate from 0 to 60 mph), and hence will not replace traditional rocket motors. They are, however, about 10 times more efficient than chemical thrusters. Ion engines may therefore become a primary technology for manoeuvring craft in deep space.

Many future robotic missions to asteroids are now planned, with most intended to analyze their surface composition. NASA also has an Asteroid Redirect Mission (ARM) that is far more ambitious. To cite the project's website:

> NASA is developing a first-ever robotic mission to visit a large near-Earth asteroid, collect a multi-ton boulder from its surface, and use it in an enhanced gravity tractor asteroid deflection demonstration. The spacecraft will then redirect the multi-ton boulder into a stable orbit around the Moon, where astronauts will explore it and return with samples in the mid-2020s.

The ARM mission 'will demonstrate planetary defence techniques to deflect dangerous asteroids and protect Earth if needed in the future'. It addition, ARM will also serve to advance both unmanned and manned deep space technologies. Already NASA has identified at least four good candidate asteroids, and plans to launch its ARM unmanned spacecraft by 2020. This will travel to the selected asteroid and capture a boulder from its surface using a robotic arm. The ARM spacecraft will then redirect this asteroid mass into a stable 'distant retrograde orbit' around the Moon. When this has happened, astronauts will travel to lunar orbit in the mid-2020s to take a closer look.

Key to the second part of the ARM mission – and quite possibly future asteroid exploration and eventually asteroid mining – is the development of the next generation of manned space transportation. In the immediate future, this will come in the form of NASA's Orion capsule – or 'multi-purpose crew vehicle (MPCV)' – which will launch atop its new Space Launch System (SLS) rocket.

Back in the last chapter I mentioned the Crew Dragon and CST-100 manned space capsules that are being developed for NASA by SpaceX and Boeing. These are intended to ferry people to and from Earth orbit, and would hence have to be altered significantly to play any role in asteroid prospecting or mining operations. In contrast, Orion is being built as a manned space capsule that will be technically capable of travelling to the Moon, to Mars, and hence potentially to an NEA. As NASA further explain, Orion:

> . . . is built to take humans farther than they've ever gone before. Orion will serve as the exploration vehicle that will carry the crew to space, provide emergency abort capability, sustain the crew during the space travel, and provide safe re-entry from deep space return velocities.

Orion accommodates four astronauts, and is being built primarily by Lockheed Martin. The first capsule orbited the Earth on a four-hour test flight in December 2014, with a further unmanned test currently planned for 2017 or 2018. This is intended to involve a cruise around the Moon, with Orion's first manned flight probably also to be a jaunt around our lonely satellite. If all goes well, the manned Orion mission to the ARM asteroid bolder will then take place in the mid 2020s.

NASA plans can and do change. The Orion capsule itself was, for example, initially developed as part of the aborted Constellation Programme that was intended to return Americans to the surface of the Moon by 2020. We therefore cannot be certain that an Orion capsule will ever carry astronauts to a chunk of asteroid material in lunar orbit. This point noted, Orion and its SLS rocket are now hardware rather than drawing board ideas. They therefore constitute one of humanity's two most likely forms of manned asteroid transportation in the next decade or so. The only currently realistic alternative would be China's Long March-9 rocket, coupled with some future version of its Shenzhou manned space capsule. I will say far more about China's space programme in the next chapter.

MINING OPERATIONS

Once sufficient prospecting has taken place, and appropriate space transportation has been developed, we will be ready for asteroid mining operations. Fortunately, with deposits of metals likely to be in a pristine condition and accessible on or close to the surface, their mining may be a relatively straight-forward operation. Or as Brad R. Blair describes in his interesting article on *The Role of Near-Earth Asteroids in Long-Term Platinum Supply*, 'mining methodologies appear to be simple, requiring the separation of finely pulverized soil in a low gravity, high vacuum environment'.

According to Mark Sonter, a mining and metallurgical consultant, we know from meteorite analysis that asteroids rich in deposits of platinum group metals are likely to contain around 100 ppm (parts per million) of pure metal. This compares to ore grades of between 5 ppm and 10 ppm in mines here on Earth, and means that there will be about 1 kilogram of metal in every 10 tonnes of asteroid material (or what is technically known as regolith).

In common with mining here on the Earth, the asteroid mining of metals will require deposit-rich regolith to be excavated and if necessary crushed. Mechanical, magnetic and other separation techniques will then need to be used to extract final raw materials. The mining of water, hydrocarbons and other volatiles will alternatively require regolith to be suitably heated to release the relevant liquids or gases, with mechanical filtering or chemical post-processing subsequently required.

In this context, it should be noted that the native temperature of an asteroid varies according to its distance from the Sun, but is typically at least as cold as -70 degrees centigrade.

Future asteroid miners will be challenged with the difficult choice of deciding where and how to engage in regolith extraction and processing. The options will be to accomplish everything 'out in the field' on the asteroid itself, or to transport an asteroid or its regolith for mining and material processing in an easier-to-access location, such as lunar orbit. The first, in situ option will demand an extreme expertise in deep space engineering, while the latter will require bulk deep space transportation systems. Opting to process and extract regolith in lunar orbit also involves the inherent disadvantage of transporting an enormous quantity of rock, most of which will end up being discarded.

Ultimately, the choice of whether to mine and process in situ or closer-to-home is likely to depend on individual asteroid location, future technology developments, the value of the materials to be extracted, and whether they are destined for extraterrestrial or terrestrial markets. When the decision is taken to do things in situ, a range of further sub-options will also present themselves.

For example, on larger asteroids with significant deposits of valuable raw materials, bases may be established. As conceptualized in figure 7.3, these would include all of the infrastructure for regolith processing. Asteroid bases could be staffed by humans, robots, or some transhuman combination thereof, with any human or transhuman quarters cut deep into the rock to provide shielding from the radiation hazard posed by cosmic rays.

Asteroid bases would need to store their processed raw materials until the asteroid's orbit brought it within a reasonable distance of the Earth. Their precious spoils would then be picked up and sent back to the motherworld or another required location, perhaps in vast 'rocket trains'.

Figure 7.3: Asteroid Mining Base. Image: Christopher Barnatt.

As an alternative to establishing long-term bases, limpet ships may temporarily attach themselves to asteroids large or small. As conceptualized in figure 7.4, on larger asteroids such mining craft may land at surface locations with high metal or other valuable raw material concentrations. They would then eat into the surface, and extract and immediately process regolith. Refined raw materials would subsequently build up in their cargo bays over a period of weeks, months or even years, with the craft periodically relocating as required. When its cargo bay was full, the limpet ship would journey back to Earth or lunar orbit to deliver its bounty.

Limpet ships may be a particularly good option for mining precious metals, small volumes of which are extremely expensive. For example, even today 1 tonne of platinum is worth in excess of $50 million. This means that a limpet ship able to accommodate 100 tonnes of platinum (which is less than 5 cubic metres in volume), would return with a cargo worth in excess of $5 billion.

As yet another in situ mining option, processing stations may be positioned near a large asteroid. Mined regolith, or

Figure 7.4: Limpet Ship Miner. Image: Christopher Barnatt.

perhaps large laser-cut asteroid sections, would then be taken from the asteroid by robotic miners and fed to the station to be refined. Once again, rocket trains or other future space haulage systems would transport away final raw materials.

Those pioneering souls hardy enough to become resident asteroid miners will be in for very long periods of duty. Transportation to and from their target asteroid will only be possible when its orbit brings it close to the Earth, and even then the trip will take months in each direction. Resident asteroid miners are therefore destined to be posted to their mines for maybe four or five years. Given that humans are not well suited to living in deep space for long periods, significant adaptations to the human form to make this possible are likely to be required. As I shall outline in the next three chapters, the pursuit of resources from space is very likely to drive the transhuman evolution of at least one future branch of our species.

Due to the potential difficulties associated with extracting and processing regolith in situ, some asteroid mining pioneers may opt to transport small asteroids for processing in lunar orbit. Asteroid processing in Earth orbit is also a technical possibility. But it is likely to be deemed far too risky to bring asteroids that close to the Earth except for exceptional purposes, such as when they are required as space elevator counterweights. Even today, lunar orbit is only a few days away by rocket.

PLANS FOR ASTEROID CAPTURE

In April 2012, a study sponsored by the Keck Institute for Space Studies (KISS) published a detailed report that investigated 'the feasibility of identifying, robotically capturing, and returning an entire Near-Earth Asteroid (NEA) to the vicinity of the Earth by the middle of the next decade'. The research was authored by participants from a wide range of bodies, including NASA's Ames Research Center and Jet Propulsion Laboratory, the California Institute of Technology, Carnegie Mellon, Harvard University and the Planetary Society. It was therefore a very serious piece of work, and noted that:

> The feasibility of an asteroid retrieval mission hinges on finding an overlap between the smallest NEAs that could be reasonably discovered and characterized and the largest NEAs that could be captured and transported in a reasonable flight time. This overlap appears to be centered on NEAs roughly 7 m in diameter corresponding to masses in the range of 250,000 kg to 1,000,000 kg.

The report went on to conclude that 'it appears feasible' to identify, capture and return a 500 tonne NEA to high lunar orbit by 2025. This, the authors suggested, would be contingent on three key developments. Of these, the first is our ability to discover and identify suitable target asteroids.

Secondly, it will be necessary to 'implement sufficiently powerful solar electric propulsion systems [or in other words ion drive engines] to enable the transportation of the captured NEA. And thirdly, we will need to have developed a sufficient lunar-faring human space presence – something that both the Americans and the Chinese are currently working toward.

The KISS report included a concept design for a flight system capable of rendezvousing with an NEA, capturing it, and using a 40 kilowatt ion drive to transport it to lunar orbit 'in a total flight time of 6 to 10 years'. Such hardware would be able to be launched into low-Earth orbit using a single, existing Atlas V-class launch vehicle, with the total cost of an asteroid capture and return mission estimated at $2.6 billion.

The mission conceptualized in the KISS report is clearly an order of magnitude more ambitious than NASA's current plan to transport a 'multi-ton boulder' of asteroid material to lunar orbit by the mid-2020s. But it could also turn out to be a critical investment for the future of humanity. Or as the KISS report expressed the matter:

> The availability of a multi-hundred-ton asteroid in lunar orbit could . . . stimulate the expansion of international cooperation in space as agencies work together to determine how to sample and process this raw material. The capture, transportation, examination, and dissection of an entire NEA would [also] provide valuable information for planetary defense activities that may someday have to deflect a much larger near-Earth object. Finally, placing a NEA in lunar orbit would provide a new capability for human exploration not seen since Apollo. Such an achievement has the potential to inspire a nation. It would be mankind's first attempt at modifying

the heavens to enable the permanent settlement of humans in space.

THE FIRST PIONEERS

While the prospect of asteroid mining may seem a distant fantasy, two companies already have pioneering ambitions. The first is Planetary Resources, which has a vision 'to do the "impossible" now'. More specifically, Planetary Resources intends to bring 'the natural resources of space within humanity's economic sphere of influence'. As the company further explains:

> Our long-term vision is nothing less than expanding humanity's resource base and extending the economy into the Solar System. Asteroids are the target to achieve this, by mining high concentrations of water and precious metals from the Near-Earth asteroids and delivering these resources to their point of need for an economic return. This mission is no less audacious than that of the East India Company in the 1600s. As was the case then, many consider the goal of asteroid mining to be too capital intensive or too long in development to be economically feasible today. We think otherwise.

Planetary Resources was initially formed in 2010 under the name Arkyd Astronautics, but then reorganized and renamed as Planetary Resources in 2012. The founders were Eric C. Anderson and Peter Diamandis, two long-standing pioneers in the commercial space business. The company employs many former NASA scientists and engineers, with notable backers and investors including film director James Cameron, signature entrepreneurs Ross Perot, Jr. and Richard Branson, and former Google CEO Eric Schmidt.

Planetary Resources also has a strategic partnership with 3D printing giant 3D Systems.

At present, Planetary Resources is focused on lowering the cost of orbital access, and developing technologies for 'commercial asteroid prospecting endeavours'. To this end it is developing a range of spacecraft called 'Arkyd'. These are small enough to 'hitch a ride' on rockets carrying another primary payload, hence enabling them to get into space cost-effectively. Current models include the ARKYD 3 (A3R) and ARKYD 6 (AR6). The first ARKYD 3 to get into space was carried to the International Space Station on a SpaceX Falcon 9 rocket in April 2015, before being released into orbit to commence a 90-day test mission.

To build its business, Planetary Resources plans to develop a largely-3D-printed orbital space telescope called the ARKYD 100. Larger craft known as the ARKYD 200 Series Interceptor and the ARKYD 300 Series Rendezvous Prospector are intended to follow, and will be capable of visiting asteroids to determine their composition. The ARKYD 300 is hoped to demonstrate a 'low-cost interplanetary capability' of interest to NASA and other organizations, and will be deployed to target asteroids in swarms that will 'distribute mission risk across several units, and allow for broad based functionality within the cluster of spacecraft'.

In January 2013, a startup called Deep Space Industries (DSI) also joined the commercial asteroid mining crusade, noting that 'the resource potential of space outstrips that of any previous frontier'. The company's founders included commercial space pioneers Bryan Versteeg, Rick Tumlinson, John S. Lewis, David Gump and Stephen Covey, a team who intend to change:

> ... the economics of the space industry by providing the technical resources, capabilities and system integration required to prospect for, harvest,

process, manufacture and market in-space resources. These resources, found on easily accessible near earth asteroids, will provide unlimited energy and supplies for a growing space economy. We will produce water, propellant, and building materials to serve growing space markets. From extending the profitability of commercial satellites, to providing life support and power to new private-sector orbiting research stations, Deep Space Industries is enabling the settlement of the final frontier.

Like Planetary Resources, DSI is focused on lowering the cost of space access, and in particular developing innovations that include nano-satellites (nanosats) and robotic and mining technologies. To date, DSI has developed its Agile Nanosat Platform, as well as a 'Mothership' carrier spacecraft that is 'designed to deliver nano-satellites to deep space targets such as the moon, asteroids, and beyond'. Like the space hardware developed by Planetary Resources, the DSI mothership and nanosats will travel into space as a secondary payload on third-party rocket launches.

In time, DSI hopes to become 'the fuel depot and resupply station in orbit' that will service the growing private and public space sector. In the shorter-term, the company has started to generate revenues from commercial space and research activities that leverage its particular innovations and expertise. Such activities have already included working with NASA on its ARM mission, and signing a contract to develop a satellite constellation for Bitcoin. In May 2015, DSI also signed two contracts with NASA that will see it develop simulated asteroid regolith for the testing of asteroid mining technologies, as well as methods for manufacturing spacecraft propellant from asteroid materials.

Looking into the future, DSI has grand plans not just to prospect asteroids, and to harvest and process their resources, but in addition to manufacture in space. To this end, the company has applied for a patent on a new 3D printing technology that it calls the MicroGravity Foundry. This 'uses a gas to extract nickel directly from the nickel-iron ore that is common in asteroids, and deposits it into 3D shapes'. In the future, this new variant of local digital manufacturing could allow space-based infrastructure – like solar power satellites – to be manufactured directly from asteroid regolith. In time, DSI also hopes that its MicroGravity Foundry will enable vast human settlements to be constructed in space.

A SERIOUS PROPOSITION?

I earn a reasonable proportion of my living talking to business audiences about the future. Sometimes the client has a very specific requirement for the content to be addressed, such as the impact of a particular new technology on healthcare, the arts or financial services. Alternatively, I am often just asked to 'shake things up' by talking about a wide range of possible future developments. This is always an exciting brief, but also a tricky one, as many people in business approach a presentation by a futurist with a fair degree of scepticism. I therefore have to tread a fine line between opening minds to new technological and other possibilities, and ensuring that I maintain at least some level of credibility with my audience.

Due to the above, asteroid mining is not a subject that I include in most of my presentations. For many in the City, the idea is simply too far removed from current reality, and hence not something which they are prepared to take seriously. 3D printing, synthetic biology, next-generation nanotech, artificial intelligence and even humanoid robots are Next Big Things that most business people are prepared to entertain as a future possibility. But mining asteroids tens of millions of miles away out in space? Or even in lunar orbit after we have sent a robot to capture a large chunk of space rock? Well, this is not a subject that can currently be broached with most mainstream business audiences.

The fact that the majority of the business world is not yet willing to entertain the idea of obtaining resources from space is both sad and quite alarming. As I argued at the start of this chapter, sometime this century we will either have to start harvesting resources from space, or will need to prepare for a mass depopulation. In the second half of this century, we will simply be unable to sustain a population of nine billion people unless we manage to bring additional external resources into the closed system of Planet Earth. The answer to the question 'is mining the asteroids a serious proposition?' therefore has to be 'yes, if we want to avoid a very unpleasant future of relentless decline'.

In December 2014, Deep Space Industries co-founder John S. Lewis published an eye-opening book called *Asteroid Mining 101*. This predicts that, once our species starts to reside in Earth orbit and beyond, resources obtained from asteroids 'could support 400 billion people living in abundance'. This precise calculation may or may not turn out to be accurate. But I would suggest that the underlying sentiment is entirely correct.

If we do not choose to mine the asteroids, then by 2100 we will need to do something equally astonishing. Few alternative options currently present themselves, although – as Yoda once revealed – 'there is another'. And that, as we shall explore in the next chapter, would be to forsake the resources of the asteroids in favour of mining the Moon.

8
MINING THE MOON

Since the dawn of recorded history, human beings have looked up to the Moon. To date, only twelve people have stood on the lunar surface and looked back at Mother Earth, with the last Apollo astronauts having departed in 1972. But now, over forty years on, there is an increasing appetite to return.

For example, in June 2015, Professor Johann-Dietrich Woerner – the new Director General of the European Space Agency (ESA) – told the BBC that he proposed the construction of a 'village on the far side of the Moon'. This, the Professor suggested, should not be just 'some houses, a church and a town hall', but a focus for the international space community and a stepping-stone for the further human exploration of the Solar System. As the ESA's Director went on to enthuse, 'the far side of the Moon is very interesting because we could have telescopes looking deep into the Universe'. There would also be the possibility of conducting 'lunar science', as well as developing and testing technologies required for any future manned mission to Mars.

Also harbouring manned lunar ambitions are a company called Space Adventures. Here the idea is to provide a commercial 'circumlunar service' that will, around 2018, begin launching missions to carry two space tourists and a Russian cosmonaut on a six-day trip around the Moon. The idea is to modify a Soyuz spacecraft to fly very wealthy individuals within 62 miles of the lunar surface, and to allow them to witness Earthrise over its horizon. If you think this sounds absolutely crazy, it should be noted that Space Adventures have already successfully organized several 10 day trips to the International Space Station (ISS), each with a $50 million price tag. Circumlunar excursions would obviously cost a great deal more, and not least because they would start with a jaunt up to the ISS. Even so, it is technically and financially plausible that Space Adventures will find at least a couple of rich takers and succeed in its lunar tourism ambitions.

Beyond European dreams and possible billion-dollar tourist plans, the Moon is starting to be seriously considered as a potential source of future energy and raw material supplies. Because of the Moon's gravitational pull, it is technically more difficult to voyage to and from its surface than that of an asteroid. On the other hand, the Moon is far closer than any other grand rock out in space. As we start to approach Peak Everything here on Earth, human beings and intelligent robots may therefore return to the airless plains of our lonely satellite with scavenging intent.

THE EIGHTH CONTINENT

In common with the asteroids, the Moon harbours substantial resources. These include cobalt, iron, gold, titanium, tungsten and the full gamut of platinum group metals. Small quantities of the nuclear fuel uranium are also thought to be available on the Moon, as is a fairly substantial supply of a rare helium isotope called helium-3. The latter may be used as the fuel in future nuclear fusion power stations – a subject to which I will return later in this chapter.

In addition to metals and potential nuclear fuels, raw Moon rock – technically known as 'lunar regolith' – may in the future be used as a space construction material. For example, lunar regolith could serve as the feedstock for

Figure 8.1: 3D Printing a Lunar Base.
Image: Christopher Barnatt.

future 3D printers that would turn it into human habitats and radiation shielding. This concept is visualized in figure 8.1, with both NASA and the European Space Agency having already conducted experiments to construct future lunar (or Martian) dwellings by spraying a binder onto successive layers of simulated regolith.

Until very recently, the Moon was believed to be a dry, desolate place. Certainly liquid water cannot persist on the lunar surface, and any water vapour would be decomposed by sunlight. Even so, in 2009, NASA's Lunar CRater Observation and Sensing Satellite (LCROSS) went in search of water on the Moon. To make this possible, NASA deployed a rocket to crash into the lunar surface, so allowing LCROSS to analyze the resultant 16 kilometre debris plume for traces of ice, hydrated materials, water vapour and other potentially valuable resources. NASA's results demonstrated 'that the lunar soil within shadowy craters is rich in useful materials', with the Moon being 'chemically active and [having] a water cycle'.

Surprisingly, most water detected was in the form of pure grains of ice. According to LCROSS principal investigator Anthony Colaprete:

> Seeing mostly pure water ice grains in the plume means water ice was somehow delivered to the Moon in the past, or chemical processes have been causing ice to accumulate in large quantities. Also, the diversity and abundance of certain materials called volatiles in the plume, suggest a variety of sources, like comets and asteroids, and an active water cycle within the lunar shadows.

There are now estimated to be about 10 billion tonnes of water ice at the lunar poles. As noted in the last chapter, water is a critical space currency, as it can keep humans hydrated, as well as being separable into the oxygen and hydrogen required to fuel rocket motors and to sustain human life.

With the discovery of water, the Moon is now regarded as an equal to many near-Earth asteroids as a space-based mining frontier. A company called Moon Express – one of the first commercial organizations with the intent to mine the Moon – has subsequently termed the Moon our 'eighth continent'. As it explains:

> We believe it's critical for humanity to become a multi-world species and that our sister world, the Moon, is an eighth continent holding vast resources than can help us enrich and secure our future. The Moon is unique in that its surface has remained relatively constant over billions of years.
>
> Most of the elements that are rare on Earth are believed to have originated from space, and are

largely on the surface of the Moon. Reaching for the Moon in a new paradigm of commercial economic endeavor is key to unlocking knowledge and resources that will help propel us into our future as a space faring species.

Moon Express is just one of several credible organizations whose current research and other activities may one day result in lunar mining. Other serious players include NASA, the China National Space Administration (CNSA), the Japanese Space eXploration Agency (JAXA), the European Space Agency (ESA), the Russian Federal Space Agency (Roscosmos), Astrobotic, Masten Space Systems, the Shackleton Energy Company (SEC) and SpaceIL. Since 2007, Google has also been sponsoring a competition called the Google Lunar XPRIZE. This is spurring many small, international teams to land robotic vehicles on the Moon, with the idea being to create another 'Apollo moment' for a generation that missed it the first time around.

Before 2025, many robotic and possibly a few human Moon landings are likely to take place, most of which will be achieved by those with lunar mining ambitions. It is far too early to gauge exactly whose plans will succeed, let alone which missions will provide signature stepping stones for the survival and advancement of our species. But the passion and drive of all of those planning second-generation lunar landings is palpable, and ought to be receiving far more mainstream attention.

THE LUNAR FRONTIER

To this point in history, only superpowers have successfully reached for the Moon. Back in 1959, the Soviet Union's Luna 2 probe made the first ever lunar impact, a feat matched by NASA with its Ranger 7 impactor on 28 July 1964. The Soviet Union again took the lead in the first Space Race on

3 February 1966, when its Luna 9 unmanned craft made a successful soft touchdown and transmitted images back to the Earth. NASA followed, with five Surveyor probes making successful soft landings between 2 June 1966 and 10 January 1968. Neil Armstrong then took the first human footstep on the Moon on 20 July 1969, with NASA's Apollo Program safely transporting twelve Americans to and from the lunar surface. Less widely remembered, Russia's Luna 16, Luna 20 and Luna 24 returned samples of lunar regolith to the Earth in 1970, 1972 and 1976.

After 1976, humanity did not send anything to the lunar surface until 14 November 2008, when India's Chandrayaan-1 orbiter ejected a probe that made a successful impact. On 10 June 2009, a Japanese probe called Kaguya (Selene) also impacted the Moon, followed on 9 October of the same year by NASA's LCROSS impactor. On 14 December 2013, China then joined the elite lunar club when its Chang'e 3 unmanned spacecraft made a soft touchdown.

In addition to the aforementioned missions, on several occasions in the past decade plans have been laid for a future lunar settlement. Perhaps most credibly, in December 2006 NASA announced its intent for a Moon base that was to be permanently staffed by 2024, and which was planned to allow for the 'maturation of in situ resource utilization'. This announcement was in line with former US President George W. Bush's *Vision for US Space Exploration* outlined in 2004, and which proposed to return Americans to the Moon by 2020. To help realize this plan, in 2005 NASA started a major lunar program called Constellation, although this was cancelled by President Obama in 2009. One survivor of Constellation was, however, the Orion space capsule. This is still under development, and is expected to carry human beings to a captured asteroid boulder in lunar orbit sometime in the 2020s.

In addition to its continued development of the Orion capsule and the Space Launch System (SLS) rocket that will carry it, in January 2014 NASA announced an initiative called Lunar Cargo Transportation and Landing by Soft Touchdown, otherwise known as Lunar CATALYST. This seeks to build on the success of the agency's Commercial Orbital Transportation Services (COTS) Program, which helped SpaceX and Orbital AKA to develop the spacecraft that now resupply the International Space Station (ISS). A subsequent program run by NASA's Commercial Crew and Cargo Program Office (C3PO) will result in SpaceX and Boeing ferrying astronauts to and from the ISS from around 2017.

According to NASA, Lunar CATALYST is intended 'to spur commercial cargo transportation capabilities to the surface of the Moon'. More specifically, the initiative is supporting 'the development of reliable and cost-effective commercial robotic lunar lander capabilities that will enable the delivery of payloads to the lunar surface'. As NASA further explains, 'such capabilities could support commercial activities on the Moon', and these may well include prospecting and mining. Indeed, as was noted in the initial Lunar CATALYST announcement:

> The Moon has scientific value and the potential to yield resources, such as water and oxygen, in relatively close proximity to Earth to help sustain deep space exploration. Commercial lunar transportation capabilities could support science and exploration objectives, such as sample returns, geophysical network deployment, resource prospecting, and technology demonstrations.

In contrast to COTS, Lunar CATALYST involves partnerships with private US companies on a no-funds-exchanged basis. Commercial space organizations therefore receive no money from NASA, but instead benefit from the technical expertise of NASA staff. Lunar CATALYST partners also obtain access to NASA test facilities, and can borrow equipment and obtain software for lunar lander development and testing. In April 2014, it was announced that NASA had signed Lunar CATALYST partnership agreements with three companies. These private lunar pioneers were Astrobotic, Masten Space Systems and Moon Express, and I will say more about their activities in a few pages time.

TOWARD A LUNAR BASE?

In addition to building Orion and working with its Lunar CATALYST partners, NASA continues to develop or commission a range of other activities that may one day result in lunar mining. For example, in May 2013, the agency announced its 'Resource Prospector' as 'the first mining expedition on another world'. The plan here is to build on the success of the LCROSS mission, and to send a robotic rover to a lunar polar region where it will 'excavate volatiles such as hydrogen, oxygen and water from the Moon'.

The current intention is to launch Resource Prospector by 2020. After its three-day journey from the Earth, the lander will make a soft-touchdown and deploy a lunar rover. As NASA explain, while the rover traverses the lunar surface:

> . . . it will use prospecting tools to search for sub-surface water, hydrogen and other volatiles. When an appropriate location is found, a drill will extract samples of the lunar regolith from as deep as one meter below the surface. The sample will be heated in an oven to determine the type and quantity of elements and compounds such as hydrogen, nitrogen, helium, methane, ammonia, hydrogen sulfide, carbon monoxide, carbon dioxide, sulfur dioxide – and most importantly, water!

Prior to deploying Resource Prospector, NASA intends to launch a low-cost CubeSat mission that will send a tiny, cereal-box-sized satellite to the Moon. Called Lunar Flashlight, this will piggyback its way into Earth orbit as a secondary payload on a rocket with another primary mission, before deploying an 80 metre square solar sail. Propelled by photons streaming from the Sun, the Lunar Flashlight will cruise to the Moon, before slowly spiralling down to within about 20 kilometres of its surface. It will then use its solar sail to reflect sunlight into shaded polar regions, which in turn will allow it to map the lunar south pole for water and volatile deposits.

In addition to committing to small scale missions like Resource Prospector and Lunar Flashlight, NASA's Emerging Space Office recently part-funded a bold study by NexGen space. Published in July 2015, this bore the gargantuan title *Economic Assessment and Systems Analysis of an Evolvable Lunar Architecture that Leverages Commercial Space Capabilities and Public-Private Partnerships*. This freely-downloadable and very detailed study was informed by the experience of many former NASA personnel. It also provides a fascinating insight into how rapidly and dramatically we could expand our lunar capabilities if we really wanted to.

The NexGen study set out to determine if America could return humans to the Moon, and establish a permanent human settlement, using commercial partnerships, and within existing NASA human spaceflight budgets. The quite startling finding was that this is indeed possible. Such a conclusion was 'based on the experience of recent NASA program innovations, such as the COTS program [which indicated that] a human return to the Moon may not be as expensive as previously thought'. Indeed, according to NexGen:

> America could lead a return of humans to the surface of the Moon within a period of 5-7 years from authority to proceed at an estimated total cost of about $10 billion (+/- 30%) for two independent and competing commercial service providers, or about $5 billion for each provider, using partnership methods.

The NexGen study proposed the establishment of an International Lunar Authority as the best mechanism for managing the commercial and technical risks involved in developing sustained lunar operations. Multiple program phases were advocated, including robot scouting missions and human sorties; the development of lunar in situ resource utilization (ISRU) capabilities for mining lunar ice and turning it into propellant; and the establishment of a large-scale lunar mining and propellant production operation that would deliver rocket fuel to lunar orbit.

The NexGen study concluded that a permanent commercial lunar base 'might substantially pay for its operations by exporting propellant to lunar orbit for sale to NASA and others to send humans to Mars'. It subsequently proposed that:

> America could lead the development of a permanent industrial base on the Moon of 4 private-sector astronauts in about 10-12 years after setting foot on the Moon that could provide 200 metric tonnes of propellant per year in lunar orbit for NASA for a total cost of about $40 Billion (+/- 30%).

In support of its conclusions, the NexGen study contains detailed technical and risk assessment proposals. These include the development of a lunar Command Module/Service Module based on a modified SpaceX Dragon V2

('Crew Dragon') spacecraft; the use of the forthcoming SpaceX Falcon Heavy and United Launch Alliance (ULA) Vulcan rockets; and the creation of a new, reusable lunar module (RLM). The latter would incorporate SpaceX's SuperDraco engines, as well as life support systems also sourced from their manned Dragon spacecraft.

The NexGen study is neither NASA nor US Government policy. Nevertheless, it includes a lot of inspirational thinking, as well as highlighting some important practicalities. These include the fact that SpaceX developed their Falcon 9 rocket and Dragon spacecraft for $443 million, or about eight times less than it would have cost NASA using its traditional methods.

CEMENTING SUPERPOWER STATUS

While today America is the only nation to have landed human beings on the Moon, the first country to return breathing beings to the lunar surface is probably going to be China. In the 1960s, America won the first Space Race in a peacetime battle for global supremacy fought against the former Union of Soviet Socialist Republics (USSR). Following the fall of the Berlin Wall in 1989, and the subsequent break-up of the USSR in 1991, for a time America reigned supreme as the world's only superpower. Yet for decades China has been on the rise, and is set to become the world's greatest superpower sometime in the 2020s (if indeed it has not already achieved this position).

Perhaps in part to cement its superpower status, and certainly to catalyze radical technological innovation, in 2004 China established the Chinese Lunar Exploration Program (CLEP). This is also known as the Chang'e Project, with the Moon Goddess Chang'e featuring prominently in Chinese legends.

The CLEP initially set itself the three milestones of 'orbiting', 'landing' and 'returning'. It met the first of these in October 2007, when its Chang'e 1 orbiter travelled to lunar orbit. There it scanned the entire Moon to generate a high definition 3D map, as well as conducting a spectral analysis of the likely distribution of potentially useful resources. A second orbital probe, Chang'e 2, followed in October 2010, travelling to the Moon more quickly than Chang'e 1 and making even more detailed observations.

CLEP Phase 2 immediately commenced, with Chang'e 3 successfully making a soft landing on the Moon on 14 December 2013. This craft carried a robotic lunar rover called Yutu (or 'Jade Rabbit'), which was equipped with cameras, near-infrared and X-ray spectrometers, and a lunar penetrating radar (LPR). The latter was able to probe about 400 metres below the lunar surface, and found evidence of at least nine distinct rock layers. This means that there has been a surprisingly high level of geological activity on the Moon over the past 3.3 billion years.

Chang'e 3 was so successful that a back-up mission (Chang'e 4) was altered to become a non-landing equipment test named Chang'e 5 T1. This launched on 31 October 2014, completed a flyby of the Moon, and returned a test capsule to the Earth which successfully touched-down in Inner Mongolia. This left the stage set for the launch of Chang'e 5 in 2017, which is intended to make a lunar landing, gather samples, and return them to the Earth.

Following the success of Chang'e 3, *People's Daily* – the newspaper of China's Communist Party – explained how the country's space program needed sustained investment to achieve 'broad scientific goals', to 'examine potential mineral and energy resources on the Moon', and to reap 'even more unexpected benefits'. Many believe that the ultimate goal of the CLEP is to set human beings on the Moon. Given that the official insignia of the program is a lunar crescent arcing around two human footprints, this also appears to be a very reasonable deduction.

In superpower terms, a manned Chinese Moon landing would match the achievement of the United States in 1969, and would even supersede Apollo if it formed part of a broader, successful plan to establish a manned lunar base or to commence lunar mining operations. In this context, it is worth noting that the CLEP's chief scientist, Ouyang Ziyuan, is a geologist and chemical cosmologist who has long advocated the exploitation of lunar resources.

As the last few pages have illustrated, so far the CLEP has proved extremely successful. In turn, this suggests that it is quite probable that China will achieve a lunar footfall sometime in the 2020s – and very possibly before any other nation or organization makes a human return to the Moon. I would indeed place a small bet that, when a team of Americans finally survey an asteroid bolder in lunar orbit, they will be able to look down on a Chinese landing craft on the lunar surface.

THE NEXT MOON RACE

In addition to the Chinese and American governments, many other parties may be involved in the next Moon race. For a start, Russia has on many occasions signalled an intent to land cosmonauts and to build a lunar base. For example, in January 2006, Nikolai Sevastyanov – then head of the Energia space corporation – said that Russia was planning to set up a permanent Moon base by 2015, with an 'industrial-scale' mining operation to be established by 2020. Six years later, there were also reports that the Russian Federal Space Agency, Roscosmos, was in talks with NASA and the European Space Agency concerning the creation of a manned research outpost. At the time, the agency's chief – Vladimir Popovkin – noted that they did not want 'to just step on the Moon', but to 'begin its exploration' and either 'set up a base on the Moon or to launch a station to orbit around it'.

In April 2014, on the 53rd anniversary of the first manned spaceflight by Yuri Gagarin, Russian Deputy premier

Dmitry Rogozin told the world's media that Russia had plans to inhabit the Moon 'forever'. While such a statement was probably no more than New Cold War bravado, in January 2015 a Russian company called Lin Industrial announced plans to build a $9.4 billion 'lunar base camp' within ten years of receiving government approval. According to the Russian News Agency TASS, this would require 13 launches of heavy rockets to ferry equipment to the Moon, followed by another 37 to establish 'adequate living conditions'. Both TASS and the Russian Academy of Sciences Institute of Space Research did, however, go on to caution that Lin Industrial's plans were probably a little premature.

The prospect of a future Russian manned presence on the Moon was again brought to the attention of the World's media in May 2015, when Chinese Vice Premier Wang Yang met with his Russian counterpart Dmitry Rogozin. Here the announcement was for a joint Chinese and Russian space exploration project that could one day lead to the establishment of a Sino-Russian Moon base. Not to be outdone, in April 2015 JAXA announced that Japan would land an unmanned rover on the Moon by 2018 or early 2019. JAXA has also set a target date of 2025 for a manned Japanese lunar landing.

THE GOOGLE LUNAR XPRIZE

Several private companies now also have more modest but potentially credible lunar ambitions. In part, such pioneers are being spurred on by the Google Lunar XPRIZE. This was launched in September 2007 to 'incentivize space entrepreneurs to create a new era of affordable access to the Moon and beyond'. Total prize funds to a maximum value of $40 million are available, and are due to be awarded to those teams 'who are able to land a privately funded rover on the Moon, travel 500 meters, and transmit back high definition video and images'. The first team that successfully achieves this requirement will be awarded the $20 million Grand Prize, with the team that comes second winning $5 million. Several bonus or 'milestone' prizes are also up for grabs, with $5.25 million having already been awarded to those teams who have developed and tested suitable landing, mobility and imaging technologies. There are also bonus prizes for surviving the lunar night or visiting an Apollo landing site.

While Google itself currently has no plans to mine the Moon, in sponsoring the XPRIZE the company is clearly trying to promote and assist the future exploitation of off-world resources. As the XPRIZE website further explains:

> The Moon is a treasure chest of rare metals and other beneficial materials that can be used here on Earth. A successful Google Lunar XPRIZE would result in cost-effective and reliable access to the Moon, allowing for the development of new methods of discovering and using space resources, and in the long-term, helping to expand human civilization into space.

Between 2007 and 2010, 29 teams from 16 countries registered for the grand challenge to create another 'Apollo moment'. By November 2015, 16 teams from 13 countries remained in the game. These included Moon Express and Astrobotic from the United States, Team Indus from India, Hakuto from Japan, and Part-Time Scientists from Germany, all of whom have received one or more milestone prize awards. All 16 teams have until 31 December 2017 to land on the Moon – a deadline that has twice been extended from an initial endpoint of 31 December 2015. The December 2017 deadline is, however, likely to be definitive, as in October 2015 Israeli team SpaceIL signed a verified launch contract. This will see its spacecraft leave the Earth in the 'second half of 2017' on a SpaceX Falcon 9 rocket purchased for a 'rideshare' mission by Spaceflight Industries.

While some Google Lunar XPRIZE contenders, such as Part-Time Scientists, are groups of engineers and entrepreneurs, others are more singular in their composition. Moon Express, for example, was founded by Bob Richards, Naveen Jain and Dr. Barney Pell in August 2010 to 'blaze a trail to the Moon to unlock its mysteries and resources for the benefit of life on earth and our future in space'. The company has already signed a Lunar CATALYST partnership agreement with NASA, and has tested a prototype lander at the Kennedy Space Center in Florida.

Also already working with NASA as part of Lunar CATALYST are XPRIZE contenders Astrobotic. To help fund its operations, the company now offer a service called MoonMail. This allows anybody to send a small memento to the Moon for safekeeping for 'centuries to come'. People who want to send a photograph, a piece of jewelry or another keepsake to the Moon as part of the 'first commercial lunar landing' are already able to sign up. Prices start from $460 for an hexagonal capsule 0.5 inches across and 0.25 inches deep. Interested parties should note that only inert, non-hazardous items will be accepted.

Whether or not any of the Google Lunar XPRIZE contenders make a successful touchdown, it seems likely that private companies will start to land on the Moon in the coming decade. A final potential candidate I would like to mention here is The Shackleton Energy Company (SEC), who one day hope to become 'just a boring space utility' who 'fuel the space frontier by supplying Earth, Moon and Mars with water, fuel and solar power'.

Shackleton Energy has already published detailed plans that it hopes will allow it to 'place a team on the Moon within 8 years', with its first revenues (from water mining) expected to be generated 'within 4 years of program start and full break-even within 12 years'. In common with many others, Shackleton Energy argues that 'much like gold opened the West, lunar water will open space like never before'. If you want to delve further into the possibilities for exploiting lunar resources, the company's website at shackletonenergy.com is well worth a visit.

CHALLENGES AHEAD

Any government or company with plans to mine the Moon is going to face a wide range of challenges. For a start, transportation to and from the Moon is likely to remain both an expensive and a technically complex endeavour. All prospective lunar miners will therefore rely on those companies – such as Arianespace, Orbital ATK, International Launch Services, the United Launch Alliance, SpaceX and Energia – who fly heavy lift rockets capable of getting people and cargo into Earth orbit and beyond.

Once people and equipment are in Earth orbit, getting them to the Moon and back is less technically problematic. Even so, transporting any quantity of mined material from the Moon to the location in which it is needed will still be no mean task. Granted, if the first lunar miners focus on extracting water and processing it into oxygen and rocket fuel, then all of their customers will be in space. But this still means that hundreds, thousands or even tens of thousands of tonnes of processed materials will have to be regularly transported into lunar orbit. This could be achieved using reusable rockets powered with fuels mined in situ on the Moon. Alternatively, a lunar space elevator (LSE) could be constructed. This would allow materials to be transported from the Moon's surface in a carriage that climbs a very long cable. A lunar space elevator may therefore one day improve the financial viability of lunar mining.

As noted in chapter 6, a terrestrial Earth-to-Earth-orbit space elevator could not be constructed with existing technologies. Terrestrial space elevators will hence only become a possibility when we are able to spin very long strands of carbon nanotubes or other as-yet-unrealized build substances. In contrast, as the Moon's gravitational pull is only about 17 per cent that of the Earth, a lunar space elevator could be constructed using existing materials. For example, a viable lunar space elevator cable could be fashioned from the very light but very strong synthetic fibers Dyneema, Zylon or Magellan-M5.

While a terrestrial space elevator cable would be held taut primarily by the Earth's rapid rotational acceleration, for a lunar equivalent the Moon's gravitational force would need to be counterbalanced by the Earth's tidal acceleration. This means that a lunar space elevator would require a very long cable indeed. In fact, lunar space elevator advocate and pioneer T.M. Eubanks has calculated that cable length would need to be around 264,000 kilometres. This is because the tether would have to extend from the lunar surface to a counterweight positioned at a balance point between the Moon and the Earth.

Beyond problems inherent in space transportation, future lunar miners will also have the considerable challenge of constructing, operating and maintaining a considerable quantity of mining infrastructure on the lunar surface. Even though lunar resources are likely to be found in a relatively pristine condition, large and complex machines will still be needed to dig out, sift, heat, chemically treat and otherwise process lunar regolith. Such machines will also have to work very reliably in extremes of heat, and while subject to very considerable dust contamination.

As the Apollo astronauts discovered, the lunar topsoil is as fine as flour, but as tough as sandpaper. Moon dust therefore poses a significant potential problem for any company wishing to operate machinery on the Moon, let alone those planning to create even more dust via mining operations.

Any lunar dust that gets carried into a human habitat could also present a considerable respiratory health risk.

Financial and engineering challenges aside, prospective lunar miners may also face a legal minefield. All of today's lunar mining hopefuls describe the Moon as a new resource frontier that is ripe for their picking. This may also be true. However, some are already questioning just who owns the Moon, and if anybody has the legal right to stake a claim on its resources.

At present, the legal framework governing the Moon is contained in an international agreement called the *Treaty on Principles Governing the Activities of States in the Exploration and Use of Outer Space, including the Moon and Other Celestial Bodies*. Otherwise known as the *Outer Space Treaty*, this limits the use of the Moon and other celestial bodies to peaceful purposes. More significantly for potential lunar miners, the treaty states that the 'use of outer space, including the Moon and other celestial bodies, shall be carried out for the benefit and in the interests of all countries, irrespective of their degree of economic or scientific development, and shall be the province of all mankind'.

As Article II of the *Outer Space Treaty* goes on to note, 'outer space, including the Moon and other celestial bodies, is not subject to national appropriation by claim of sovereignty, by means of use or occupation, or by any other means'. This implies that any country which establishes a lunar mining operation will not actually own any resources that they subsequently recover and refine. It may therefore turn out to be very significant that the *Outer Space Treaty* makes no mention of any potential commercial ownership of lunar resources. Many consider this omission to be an indication that any company is free to claim lunar resources. But this may not turn out to be a legally defensible position.

The *Outer Space Treaty* was first opened for signature by the United States, the Soviet Union and the United Kingdom in 1967, and is maintained by the United Nations Office for Outer Space Affairs. As of May 2013, 103 countries were party to the treaty, with another 26 having signed but not completed ratification.

Given that both technological developments and resource scarcity are likely to make obtaining resources from space an increasingly feasible and profitable proposition, I suspect that the *Outer Space Treaty* will be subject to significant review, extension or clarification in the next decade. The 2015 call by NexGen to establish an International Lunar Authority for managing lunar operations may hence turn out to be very well timed. It would indeed be shameful if investment in lunar exploration stalled due to legal uncertaintities concerning future resource ownership.

MOON POWER

The commercial exploitation of most lunar resources will demand significant innovations in many areas of space science and technology. There is, however, one potentially very valuable lunar material that could be economic to mine even with existing knowledge and space hardware. The substance in question is an isotope of helium called helium-3, which may fuel future nuclear fusion power stations. To understand the extraordinary possibilities associated with mining helium-3 on the Moon, we therefore need to delve into a little nuclear physics.

At present, all nuclear power stations use a process called nuclear fission to generate heat and subsequently electricity. More specifically, as shown in figure 8.2, inside current fission reactors atoms of uranium or plutonium are struck by neutrons that break them apart. This releases atomic energy in the form of heat, which is then used to produce steam, turn a turbine, and generate electricity. Unfortunately, the nuclear fission reaction also results in the production of highly radioactive nuclear waste.

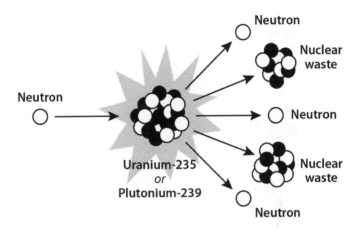

Figure 8.2: Nuclear Fission.

In contrast, in future nuclear fusion power plants, two fuels could be atomically fused together in a reaction that would release atomic energy without creating significant nuclear waste. As illustrated in figure 8.3, in current test nuclear fusion reactors the hydrogen isotopes tritium and deuterium are used as the fuels, with atomic energy released when their nuclei fuse to create helium and a neutron. This said, the so-termed 'fast' neutrons so released lead to a significant energy loss and are extremely difficult to safely contain inside the reactor. For this and a host of other reasons, scientists and engineers have so far failed to create nuclear fusion reactors that can provide a viable source of nuclear energy.

One potential solution may be to use helium-3 and deuterium as the fuels in future 'aneutronic' (power without neutrons) fusion reactors. As shown in figure 8.4, here the involved nuclear reaction results in normal helium and a proton, with the latter being relatively easy to contain magnetically within the reactor as a proton is a charged particle. Deuterium – also known as 'heavy hydrogen' – is easy to

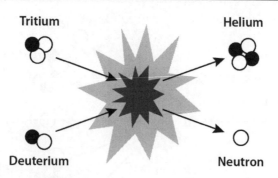

Figure 8.3: Nuclear Fusion.

obtain, as it occurs naturally in seawater. In contrast, on the Earth helium-3 is very rare indeed, with only a few kilograms produced each year as a by-product of maintaining nuclear weapons. Helium-3 is, however, emitted by the Sun, and is carried out into the Solar System by its solar winds. Sadly our home planet's atmosphere and magnetic field prevent any of this precious gas arriving on the surface of the Earth. But as the Moon does not possess such impediments, the lunar regolith has been absorbing helium-3 for billions of years.

Broad estimates suggest that there are around a million tonnes of helium-3 stored in the first few metres of lunar dust (or in other words in those layers of powdered regolith that have been stirred up by asteroid bombardment). Chinese scientists Wenzhe Fa and Ya-Qiu Jin have made more precise calculations, with their lunar inventory of helium-3 including about 650,000 tonnes on the lunar nearside, and 278,000 tonnes on the lunar farside.

As a future nuclear fusion fuel, it has been estimated that 1 tonne of helium-3 could deliver the equivalent energy of about 50 million barrels of crude oil. In turn, helium-3 has been estimated to be worth up to $3 billion a tonne. Further estimates suggest that between 25 and 40 tonnes of helium-3 could fuel

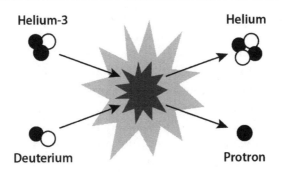

Figure 8.4: Aneutronic Nuclear Fusion Using Helium-3.

the United States for a year, with somewhere between 100 and 150 tonnes potentially being sufficient to meet the annual power requirement of the entire human race.

Mining helium-3 on the Moon would require vast areas of the lunar surface to be dug up, with the excavated regolith then subject to an extraction process that would include heating it to around 600 degrees centigrade. Conducting such industrial-scale mining and material processing operations on the Moon would clearly present an enormous challenge. The creation of a new generation of aneutronic fusion power stations to consume the fuel here on the Earth would also hardly be child's play. This said, transporting 100 or so tonnes of helium-3 from the lunar surface to the Earth each year is already not an unrealistic proposition. Later this century, lunar helium-3 mining may therefore kickstart the Resources from Space Revolution.

At various moments in history, the United States, Russia and China have all linked lunar exploration announcements to helium-3 prospecting or even future mining. Most recently, China's lunar ambitions have been strongly associated with obtaining helium-3, and with very good reason.

As it continues to industrialize, China will need to source a lot more energy, while at the same time seeking to rid itself of its current pollution problems. At its current rate of growth, even if China were to burn all of the fossil fuels extracted from the Earth each year, sometime in the 2030s it would start to face serious energy shortages. China therefore has a very great incentive to develop lunar transportation and mining technologies that could allow it to solve its future energy and pollution problems, take significant climate change mitigation measures, and potentially become the supplier of clean energy for the entire world.

Alone as a superpower, China is the only regime to have a political system that can deliver highly expensive, long-term projects. As political analyst Fabrizio Bozzato has explained, travelling to the Moon to harvest helium-3 is also 'synergistically compatible with and reflective of the values and ambitions of President Xi Jinping'. Not long after his inauguration, the new president established a national mantra known as the 'Chinese Dream'. This also includes a 'Space Dream' as 'an important part of the dream of a strong nation'.

Looked at in the context of potential helium-3 mining, China's Space Dream and lunar exploration program may be seen in a whole new light. The fact that the lunar exploration program's chief scientist, Ouyang Ziyuan, is a geologist and an advocate of helium-3 mining really is not something to be ignored.

Rather than transporting helium-3 to the Earth to fuel terrestrial fusion power plants, it is remotely possible that the fusion plants could instead be built on the Moon. Their energy would then be transmitted to the Earth using the wireless power transmission (WPT) technologies discussed in chapter 6. Energy power loss during transmission would be a very major issue, though perhaps a forfeit worth paying to avoid the problems inherent in transporting physical resources across 400,000 kilometres of space.

As yet another alternative, power could be generated on the Moon using vast solar arrays manufactured from lunar materials and spread across great swathes of the lunar surface. Once again, the energy so harvested would be transmitted to the Earth using either microwave or laser systems.

Since the early 1980s, the development of surface-based lunar solar power (LSP) has been advocated by Professor David Criswell of the University of Houston. As he explains on his website lunarsolarpower.org, a viable LSP system would consist 'of bases on the two sides of the Moon as seen from Earth and receivers distributed about the Earth'. Satellite stations in Earth orbit may also 'receive power beams and redirect them to receivers on Earth that cannot view the Moon directly'. To avoid a power outage during the period of a new Moon or eclipse, solar reflectors may additionally be required in lunar orbit to maintain illumination on the solar arrays on its surface.

Given the scale of infrastructure required to establish a viable LSP system, almost all hardware would have to be manufactured on the Moon from materials mined and processed in situ. This may also turn out to be an effective means of utilizing lunar resources, as once again they would not have to be transported into lunar orbit, across space, and down onto another planetary body.

Should either lunar helium-3 power plants or solar arrays ever be constructed, the Moon itself – rather than any Earth-based nation – may become the next, great superpower. Those who control resources – physical or intellectual – have always reigned supreme. In the 22nd century, governmental politics may therefore end up being conducted at the level of the celestial body rather than the Earth-based nation.

FROM CONSUMING LESS TO FINDING MORE

Across the last three chapters we have explored those visions and endeavours that will progress our civilization beyond sustainability. Today, in the face of Peak Everything, there is a growing concern that we need to constrain our consumption of natural resources. In the short-term, eking out the remaining supplies of our motherworld is our only credible survival strategy. But in the longer term, we will need to transition from a mantra of *consuming less* to one of *finding more* if we are to avoid mass depopulation and relentless decline.

In the past century our species has evolved its collective mental and physical capabilities at an extraordinary rate. Not least, in little more than a century, we have taken to the skies, split the atom, invented inorganic computing, and wired our brains together with the Internet. I mention these relatively recent and quite astounding accomplishments to highlight how, in the next few decades, it is perfectly reasonable to deduce that we will make further and even more extraordinary progress. Today, obtaining resources from space may seem like a fantasy. But by 2075 or even 2050, teenagers may have no direct knowledge of a time when all resources were sourced from Planet Earth.

The first human beings to travel to the Moon made their voyage using less computer power than contained in a single smartphone. Many people may therefore reasonably wonder how and why we managed to achieve so much in space in the 1960s and 1970s, and yet apparently so little in the decades since that time. To some extent, our reason for abandoning the human push beyond orbit is directly associated with the end of the Cold War. It was, after all, superpower politics rather than economics or a survival instinct that catalyzed the first Space Race.

The above point noted, my own belief is that our transition into a space-faring civilization can only really be understood in evolutionary terms. Looked at from such a Darwinian perspective, our advancement in the past few decades has been substantial. It is just that, rather than undertaking further space excursions, we have been developing new capabilities that will fairly soon facilitate an extraordinary period of extraterrestrial progress.

Life survives by evolving to conquer new realms. Given the vast resource base waiting for us out in the Solar System, it is therefore only logical that humanity will evolve into space. To do so, we will undoubtedly need new space travel technologies. But in addition, I suspect that at least some of us will also require re-engineered bodies.

The current human form is about as suited to living and working in space as a fish is adapted to residing on dry land. In the short term, some humans may continue to venture successfully into space by taking an atmosphere with them. But, further into the future, it would be rather surprising if our civilization did not develop one or more new physical forms more suited to extraterrestrial living. Or as I argued in my book *25 Things You Need to Know About the Future*:

> When our ancestors crawled out of the oceans they had to evolve to cope and then thrive in a new environment. We should therefore not necessarily believe that humans in their current form will become tomorrow's most successful space mariners. The evolutionary leap from our first planet to the vacuum of space is, after all, at least as great as that from water to dry land.

Since Neil Armstrong took our collective giant leap, a great deal of progress has been made in the fields of AI, robotics and local digital manufacturing. Indeed, as we explored in the first five chapters of this book, many developments in these areas are soon set to burn brightly as the Next Big Thing. Looking further ahead, it is also the progress that we continue to accrue in AI, robotics and local digital manufacturing that will facilitate our evolutionary journey in pursuit of resources from space.

The first pioneers to go in search of extraterrestrial riches are likely to be robots with silicon brains. But in time, such entirely synthetic intelligent entities will be joined on the final frontier by new versions of ourselves. Some of these beings may be entirely organic, but with a biology reprogrammed to be more suited to live in zero gravity or to survive extraterrestrial radiation. Alternatively, other future space pioneers may be a cybernetic mashup of 'natural' biology and inorganic technology that will be fuelled solely by electrical power.

We need to remember that *Homo sapiens* is not a static, finished creation. As we continue to evolve, at least some of our descendants – and maybe even some of those people alive today – are subsequently destined to become 'transhumans'. A proportion of these next-generation citizens will be off-world optimized. Meanwhile others may be re-engineered to possess a greater intelligence, an improved physical stamina, an extended life span, or perhaps simply to aesthetically conform to parental or individual whim.

Without doubt, the organic and inorganic technologies that will soon allow a radical alteration of the current human form are astounding. Our final two chapters will therefore explore the alternatives that lie ahead for proactive human evolution.

PART IV

TRANSHUMAN EVOLUTION

9

POST-GENOMIC MEDICINE

At a launch event held in September 2015, Apple revealed its iPhone 6s and iPhone 6s Plus. These latest smartphones offered exciting new features including 3D touch, higher resolution cameras and faster processors. Followers of the global religion created by Steve Jobs were subsequently delighted, and soon began worshiping with their credit cards to obtain the new hardware.

In common with the members of other technology tribes, most Apple fans have developed an obsessive desire to regularly upgrade. Indeed, whether they are loyal to Apple, Microsoft or Google, many people now appear unsatisfied with computer hardware or software that is more than a few years old. Even beyond the high technology sector, regular upgrades of our most utilized possessions have come to be expected. Except, that is, when it comes to the hardware and software of ourselves.

A minority of the population do now spend a great deal of time and money attempting to improve their physical and mental fitness. This point noted, even the best diet and exercise programs promise little more than to raise the state of the body to an optimal condition and to try and keep it there. Beyond the age of 40, most people are, in fact, reasonably content if a fitness regime can just slow the 'inevitable' degeneration associated with the ageing process.

In common with its patients, the healthcare sector remains almost exclusively focused on medical maintenance. Granted, some clinics do now offer cosmetic surgeries. But beyond nip-and-tuck or silicon insertion, right now it is simply not possible for a healthy person to visit their doctor to receive a 'generational upgrade'. Prospective parents also have to settle for the delivery of offspring in their own image, rather than babies with features that they themselves do not possess. It is indeed rather curious that we have built a society that expects the constant upgrade of almost all physical items except those fabricated from human flesh and bone.

THE TRANSHUMAN AGENDA

While we currently accept the human body as a finished creation that cannot be upgraded, this is unlikely to remain the case for that much longer. This is because local digital manufacturing, AI and humanoid robots will soon allow the conscious augmentation and 'improvement' of *Homo sapiens*. Over the next few decades, at least some people are therefore destined to consciously upgrade their birth hardware in order to become 'transhuman'.

The words 'transhuman' and 'transhumanism' were first coined by biologist Julian Huxley in 1927. In his book *Religion without Revelation*, Huxley noted that the human species may choose to 'transcend itself', and that a new name was needed for such a belief. As he went on to ponder, 'perhaps transhumanism will service: man remaining man, but transcending himself by realizing new possibilities'.

One of the conceptual gurus of transhumanism is Max More, who in 1990 wrote an essay entitled *Transhumanism: Toward a Futurist Philosophy*. This is often regarded as the foundation of modern transhumanist thought, and defines transhumanism as 'a class of philosophies of life that seek the continuation and acceleration of the evolution of intelligent life beyond its currently human form and human limitations by means of science and technology'.

Today, an international organization called Humanity+ (formerly the World Transhumanist Association), provides a focal point for transhumanist activity. Over on its website at Humanityplus.org, the movement describes itself as 'an international nonprofit membership organization that advocates the ethical use of technology to expand human capacities'. Or as Humanity+ goes on to explain, 'we want people to be better than well'.

Humanity+ adopted and maintains the *Transhumanist Declaration*. This was drawn up by an international group of contributors in 1998, and notes that humanity now stands on the brink of being able to overcome 'aging, cognitive shortcomings, involuntary suffering, and our confinement to planet Earth'. The declaration goes on to state that 'humanity's potential is still mostly unrealized', and that it favours individuals having a 'wide personal choice over how they enable their lives'. As point 8 of the declaration then makes explicit, this may include the use of many current and future 'human modification and enhancement technologies'.

Transhumanism may become a significant future philosophy that will serve to proactively transition healthcare from its present focus on medical maintenance, and toward the additional practice of providing upgrade options. Almost inevitably, transhumanism has already been likened to 'playing at god', with ethical and religious concerns raised in some quarters. In this context, it is worth noting that, since 1900, modern medicine has doubled the average life span, with the majority of human beings now spending a period of time alive that is 'unnaturally long'. It could therefore be argued that transhumanism is simply seeking to build on our current success in achieving a technological life extension.

As Humanity+ argue in their mission statement, the Western world's consensus on what is 'normal' has established certain precedents, and these have 'not kept up with the advances in technology or science'. As they further contend, 'external devices' – including smartphones, smartwatches and wearable bio monitors – are all expanding human capabilities, with some medical advancements having already 'broken through the glass ceiling on biological determinism'. Whether some people like it or not, we therefore now stand on the brink of a Brave New World in which individual and quite possibly species evolution is going to fall under our conscious control.

At a fundamental, practical level, there are three distinct ways in which transhumanism may be pursued. Most straight-forwardly today, the first of these is to take all possible steps to enhance individual health and long life. Many transhumanists therefore engage in physical fitness activities, such as aerobics, yoga and pilates, as well as practising meditation.

Transhumanist measures currently adopted also include making changes to what goes into the body, for example by eating a calorie restriction and optimal nutrition (CRON) diet. This reduces calorie intake to a level 20 to 40 per cent lower than typical, while still providing necessary vitamins and nutrients. Already research has shown that CRON diets lessen the degeneration of stem cells and so reduce the incidence of cancer. Some studies in animals have even indicated that a CRON diet may extend life span by up to 40 per cent.

Beyond adopting specialist diet and fitness regimes, transhumanism may in the future involve a cybernetic fusion of the human body with inorganic technology. This option opens up a very wide range of potential possibilities, and is the subject of our final chapter. In the remainder of this chapter, we will therefore focus on the third transhumanism alternative, which is to genetically reprogram our 'natural' biological hardware.

THE BOOK OF LIFE

Like all other animals, human beings are complex biological machines whose fabrication, operation and internal maintenance is dictated by a digital code. This particular code is referred to as the 'human genome', with a copy stored in the DNA chemical sequence that resides within almost every human cell.

Throughout most of human history, doctors were unaware that DNA existed. It was, in fact, not until 1869 that Swiss chemist Johann Friedrich Miescher extracted a substance then called nuclein from white blood cells. By the early 1900s, nuclein was being referred to as deoxyribonucleic acid (DNA), and was also known to contain the code of life written in four chemical 'bases' called adenine (A), thymine (T), guanine (G) and cytosine (C). In 1944 a team lead by Oswald Avery at Rockefeller University first demonstrated that genes consisted of distinct portions of DNA. The double-helix structure of DNA was then determined by James Watson and Francis Crick in 1953.

The human genome contains approximately 3 billion chemical digits. By the last decades of the 20th century, it was becoming increasingly obvious that if doctors could learn to read, understand and manipulate this vast digital sequence, then the practice of medicine could be transformed. In 1990, a publicly-funded, international partnership known as the Human Genome Project subsequently set out to 'sequence' human DNA, or in other words to determine the exact order of its chemical bases. In 1998, a private company called Celera Genomics set itself the same goal. While initially this created some rivalry, the two research teams actually ended up working together. In February 2001, the International Human Genome Sequencing Consortium subsequently published the 'first draft' of the human genome. This was about 90 per cent complete, and was followed by the publication of the complete human genome in April 2003.

The completion of the Human Genome Project was one of the greatest achievements in human history. The 'book of life' it provided does, after all, constitute the CAD file for *Homo sapiens*, and potentially has multiple uses. Or as Francis Collins, the director of the National Human Genome Research Institute in the United States, noted when the first draft sequence was published:

> It's a history book – a narrative of the journey of our species through time. It's a shop manual, with an incredibly detailed blueprint for building every human cell. And it's a transformative textbook of medicine, with insights that will give healthcare providers immense new powers to treat, prevent and cure disease.

In addition to sequencing our entire genome, the Human Genome Project revealed several things that its researchers did not expect to discover. For a start, exactly what constituted a gene came into question. We also learned that the 'expression' of a gene – or in other words whether it is turned 'on' or 'off' – is at least as important as gene composition. Further, an understanding was cemented that every human genome is different, and that the differences between them really matter a great deal.

Impressive as it was, the completion of the human genome project was more of a beginning than an ending. Still today, the vast majority of medical practice remains 'pre-genomic', with most diagnosis reliant on external physical examinations, chemical tests, X-ray and other scans, or physically invasive investigations. As in ancient times, most treatments are then reliant on the macro-scale repair or removal of diseased or damaged tissue, or the ingestion or injection of standardized chemicals. In some respects, the practice of medicine really has not advanced a great deal for hundreds of years. But thanks to the Human Genome Project, we do now stand on the brink of a new, 'post-genomic' medical age that will deliver four highly important advancements.

Firstly, as post-genomic medicine takes hold, so genetic testing is increasingly going to be used to detect actual or potential disease. Secondly, the prescription of traditional medications will be transformed, with 'pharmacogenomics' being utilized to select drugs based on each patient's genetic profile. Thirdly, and looking further ahead, genetic therapies will become available, with doctors and AIs able to 'correct' errors in a patient's genetic code in order to cure conditions such as cancer.

The fourth and final aspect of the post-genomic medical revolution will see the pursuit of transhumanist goals. Already, this is starting to allow parents to select the characteristics of their offspring. Further into the future, it will permit 'healthy' individuals to get their bodies 'upgraded' via genetic manipulation and synthetic biology. As the AI systems required to deliver such services become available, post-genomic medicine may accelerate human evolution in a whole host of directions. The possibilities really are quite staggering. Though to understand them, we first need to step back to consider the new forms of medical maintenance that post-genomic medicine is destined to deliver.

TOWARD MOBILE DIAGNOSIS

The Human Genome Project revealed that the 3 billion chemical bases in human DNA are arranged into about 20,500 individual genes. Identifying these specific portions of DNA was a stunning achievement. Yet to use this knowledge in medical practice also requires doctors or their AIs to determine the singular or collective function of each individual gene. This will allow patients to be tested – or 'genetically screened' – for specific genetic traits, such as a predisposition to a particular disease. Just as a computer scientist will review reams of computer code to find the bugs in a program, so future doctors and AIs will routinely examine the genes (lines of genetic code) in a patient's DNA to identify actual or potential problems.

Genetic tests typically analyze a sample of a person's blood, saliva, hair or skin, or else the amniotic fluid that surrounds a foetus when it is growing in the womb. Already there are over 2,000 genetic tests that can aid in the diagnosis of over 1,000 genetic diseases. The first ever genetic tests identified mutations in a single gene that caused rare, inherited disorders such as cystic fibrosis. However, some of the more modern tests involve a far more complex analysis of a number of genes whose amalgamated makeup and expression can identify the risk of developing a condition such as cancer or heart disease.

Without doubt, it will be advancements in AI that will catalyze the innovation of a wider and more complex range of genetic tests. This is because only sophisticated, artificial neural networks will be able to determine the complex patterns of multiple gene interrelationships on which most future genetic tests will be based. As we saw in chapter 4, AI is already starting to transform medical diagnosis, and thanks to post-genomic medicine this trend is set to accelerate.

In addition to AI, developments in nanotechnology are also going to make genetic medical diagnosis more widely available. Working toward the goal of 'decentralizing, mobilizing and personalizing healthcare', a company called Nanobiosym has already developed a functional 'portable nanotechnology medical diagnosis platform' called Gene-RADAR. This is able to 'rapidly and accurately detect genetic fingerprints from any biological organism, [so] empowering people worldwide with rapid, affordable and portable diagnostic information about their own health'.

Already pre-commercial, note-book sized devices based on the Gene-RADAR version 1.0 platform can perform a diagnosis on a sample of blood or saliva up to 100 times faster than traditional methods, and for a fraction of the price. For example, an HIV viral load test – which measures the amount of HIV genetic material in a blood sample – can be conducted in a few minutes for a few dollars. This compares to a timescale of a few weeks in a developed nation – or up to six months in parts of Africa – with the current, 'traditional' price tag being around $200. Next generation Gene-RADAR platforms are intended to deliver the same diagnostic capabilities in smartphone-sized devices, wearables, and even ingestible hardware.

There is still a long way to go. But it does look increasingly certain that, in the age of post-genomic healthcare, medical diagnosis based on specific genetic tests is sooner or later going to be taking place not only in hospitals, but also in doctor's offices, patient's homes, and even inside patient's bodies on a close-to-real-time basis.

Reflecting this very probable future, Nanobiosym Gene-RADAR is just one of the contenders in the Qualcom Tricorder XPRIZE. This $10 million competition is intended to 'empower personal healthcare' by stimulating the 'innovation and integration of precision diagnostic technologies' in a portable device that will help consumers to 'make their own reliable health diagnoses anywhere, anytime'.

In 20 years, and maybe in 10, hardware like the 'tricorder' medical scanners in *Star Trek* will exist. Such next-generation medical gadgets will allow a wide range of common medical conditions to be identified on-the-spot by integrating local genetic testing with measurements of traditional diagnostics including blood pressure, respiratory rate and temperature. Quite how this will transform the relationship between patients, doctors and healthcare providers we simply do not know. It is, however, very reasonable to suggest that the implications will be very significant. The impact is also most likely to be felt by those four billion people who currently have little or no access to modern healthcare.

Even today there is a multimillion dollar industry that allows members of the public to order 'click-and-spit' genetic tests over the Internet. For example, visit 23andMe.com, and you can be supplied with a simple pack that allows a saliva sample to be returned in a pre-paid box. The included DNA will then be analyzed to provide an online report on over 100 genetically-linked health conditions and traits. As 23andMe explain, their 'saliva-based personal genome service' allows you to:

> Find out how your genetics relate to things like abnormal blood clotting, cystic fibrosis or response to certain medications. You can also see if your body metabolises caffeine quickly or if you're likely lactose intolerant. We believe the more you know about your DNA, the more you know about yourself.

23andMe go on to suggest that the results of a person's genetic analysis will enable them to 'plan for the future', 'stay one step ahead' and 'engage in [their] healthcare'. Hardly surprisingly, not everybody agrees. Most notably, the Food and Drug Administration (FDA) required 23andMe's service to be suspended in the United States between November 2013 and October 2015 due to a lack of supportive research. Even today, only customers in the United Kingdom may invest £125 to access the entire spectrum of genetic evaluations that 23andMe has on offer.

Right now, customer-facing, click-and-spit diagnostic genetic testing is probably not ready for mainstream application, with some comparative reports suggesting that results vary too widely between different online providers. This said, while current regulatory disputes and inaccuracies are undoubtedly significant, they are not insurmountable. Sometime in the 2020s, I would therefore expect click-and-spit genetic testing to go prime time. Future virtual assistants – quite possibly including Microsoft's Cortana, Google Now, Apple's Siri and Facebook's M – will subsequently be able to provide valuable healthcare advice and diagnosis by correlating personal genomic test results with a wealth of online genetic Big Data resources.

THE PHARMACOGENOMIC REVOLUTION

However sophisticated, mobile, cheap and widely available genetic testing may become, it can only ever form the first part of an effective healthcare process. Treatment, in addition to diagnosis, is after all what most patients will continue to demand. So just how and when will post-genomic innovations result in new forms of medical intervention?

As we shall see in a few pages time, several decades from now doctors and AIs will start to treat conditions like cancer by correcting 'defects' in a patient's genetic code. Yet long before this occurs, an exciting new development called pharmacogenomics – also termed pharmacogenetics – is going to transform at least some aspects of healthcare. Improvements in patient outcomes as a result of pharmacogenomics are, in fact, already starting to be reaped.

Today, when a doctor prescribes a medication, they often do so in relative ignorance of how exactly it may treat their patient. If this sounds alarming, then I must point out that doctors do the best that they can. Before drugs get to market they are subject to rigorous testing to try and ensure that they have minimal side effects and will successfully treat the condition for which they are prescribed. Even so, many patients continue to be given medications that at best do not work, and which at worst result in a sometimes fatal adverse drug response (ADR). In the United States alone, ADRs are annually responsible for over 1 million hospitalizations and more than 100,000 deaths.

Pharmacogenomics seeks to improve both drug effectiveness and drug safety by making use of individual genetic information. This means that existing, traditional drugs are still prescribed, but are selected in accordance with a genome-wide analysis of a patient's DNA. Or as a recent article in *The Pharmacist* alternatively explains, pharmacogenomics uses genetic information to provide information 'on the likelihood of drug efficacy, drug-drug interactions and adverse effects' based on hereditary (genetic) factors.

Since the dawn of modern medicine, it has been known that different people react differently when given the same medication. In many instances, this continues to make the prescription of certain drugs a bit of a lottery. For example, the drug Herceptin only works effectively as a treatment for breast cancer in about 40 percent of patients.

Until very recently doctors had to rely on intuition and trial and error when making a Herceptin prescription. This was clearly bad news for many patients, as well as being a waste of precious healthcare resources given that a course of Herceptin may cost $10,000 a month. But, by testing for an over-expression of a gene called HER-2, doctors are now able to prescribe Herceptin only to those patients for whom it is most likely to prove effective.

Already an increasing range of pharmacogenomic tests are becoming available. Assurex, for example, offer a genetic test called GeneSight that healthcare providers may use to help determine the best drugs to prescribe for patients with conditions that include depression, anxiety, chronic pain, post-traumatic stress disorder (PTSD), or attention deficit hyperactivity disorder (ADHD). As the company further explains:

> Because genes influence the way a person's body responds to specific medications, the medications may not work the same for everyone. Using DNA gathered with a simple cheek swab, GeneSight

analyzes a patient's genes and provides individualized information to help healthcare providers select medications that better match their patient's genes. Multiple clinical studies have shown that when clinicians used GeneSight to help guide treatment decisions, patients were up to twice as likely to respond to the selected medication.

The use of pharmacogenomic testing to help improve the prescription of a very wide range of drugs could be revolutionary. Not least, it should speed drug development and approval, as well as making possible traditional medications that serve only a niche population.

At present, all new medications have to be developed, tested and approved based on their suitability for the entire population. Frequently this causes very promising drugs to be abandoned, with the best treatments for some patients never making it to market due to the adverse reactions they would trigger in others. One of the really great promises of pharmacogenomics is hence to improve not just the effectiveness and safety of drugs, but also the range and quality of medications available. In turn, many pharmaceutical companies may discover that much of their abandoned research back catalogue becomes worthy of re-evaluation. There may well be tens of thousands of drugs that never made it out of testing, but which will soon become staple prescriptions for those with a particular genetic profile.

LOW COST SEQUENCING

For pharmacogenomics to really enter the mainstream, the rapid and low-cost sequencing not just of a few genes, but of entire human genomes, will first need to become very widely available. For tests that require only part of a genome to be analyzed, so termed 'gene expression microarrays' – alternatively known as 'gene chips' – may also provide at least part of the answer.

Currently in development and on sale from companies including Affymetrix and Illumina, gene expression microarrays are sensors that feature a tiny glass grid. During a genetic test, a sample of patient cells is squirted on to the microarray. This causes some of its squares to illuminate, so revealing the 'expression' (or level of activation) of particular genes. By analyzing the microarray under a microscope, it is then possible to determine which drugs are most suited to a patient's genetic profile.

When it comes to sequencing entire genomes, highly sophisticated and expensive hardware is currently required. It took the researchers involved in the Human Genome Project about 13 years to sequence the first complete human genome, with the cost estimated at around $3 billion. Clearly such a price tag would make full-genome-based pharmacogenomics a non-starter. But fortunately, just as Intel founder Gordon Moore correctly predicted that computers would increase in power and decrease in cost exponentially, so a parallel observation by economist Rob Carlson holds true for the speed and cost of genome sequencing.

Figure 9.1 illustrates the 'Carlson Curve' that reveals the exponentially falling cost of reading a complete human genome. Looking at the graph – and noting that its vertical (Y) axis has an exponential scale – it really is astonishing to realize that the price of sequencing an entire human genome has fallen from about $3 billion for the first human genome, to around about $100 million for work commenced in 2001, to the milestone figure of $1,000 or less today.

The first commercial '$1,000 genome' was achieved by Illumina in January 2014 when they launched their HiSeq X Ten System. This remains the market leading sequencing hardware, with a single machine capable of reading in excess of 18,000 complete human genomes every year. This is an average of about 50 complete genomes a day, or more than two an hour. It has to be appreciated that the HiSeq X Ten

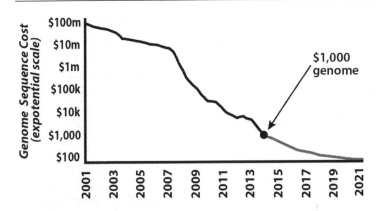

Figure 9.1: The Carlson Curve.

sequences multiple genomes in parallel. But cycle time is still less than three days, which means that entire individual genomes are sequenced in less than 72 hours.

At this rate of progress, sometime in the 2020s it should become possible to sequence an individual's entire genome for less than a dollar in a matter of minutes. This may allow a doctor – or a patient armed with a tricorder-style device like that being developed by Nanobiosym – to sequence a genome pretty much on the fly, and to prescribe medications accordingly. With the development of suitable pharmaceutical microfabricators, it may even become possible for a single piece of hardware to determine the best medication for a patient's ailment and to print out appropriate drugs.

Once a reasonable proportion of a population's genomes have been sequenced, all such information could be amalgamated into an extremely potent Big Data resource. Such an online databank could allow medical AI systems to plan and provide healthcare in a manner that would maximize patient outcomes and minimize cost. Without doubt this kind of development is going to raise all kinds of privacy and ethical concerns. Yet, as we saw in chapter 4, the potential benefit of 'data driven medicine' may be extremely significant at both an individual and a population level. In return for an increased life span and an improved quality of life, I therefore suspect that the majority of the world's population will opt-in to the crowdsourcing of their genomic information.

TOWARD GENE THERAPY

In time, post-genomic medicine will enable patients to be treated via genetic manipulation. Such 'gene therapy' is based on the principle of identifying and then correcting genetic defects in order to restore health to a patient. At present, all gene therapy is experimental and limited to around 1,000 clinical and pre-clinical trials. Not least, this is because optimal mechanisms for manipulating genes are far from being fully refined, with researchers still trying to ascertain the best methods for manipulating human DNA in order to consistently and safely achieve the desired results.

Today, four techniques are used to conduct gene therapy experiments. Firstly, an additional healthy gene may be inserted into a patient's genome in order to take on the function of a missing or inactive gene. Secondly, an unhealthy (mutated) gene may be replaced with a healthy copy. Thirdly, so-termed 'selective reverse mutation' may be used to return an unhealthy gene back to a healthy state. And finally, the expression (or level of activation) of a gene may be turned 'on' or 'off' in order to prevent a defective gene from causing problems, or to bring a healthy gene back into action. Written end-to-end in this paragraph, all of these things may sound quite straight-forward. But in the lab, let alone in a routine healthcare environment, they are currently very challenging propositions.

Most problematically, the insertion of fresh genetic material into a human cell is very hard to directly achieve. Gene therapy therefore tends to involve the creation of carriers called 'vectors' that are used to indirectly deliver therapeutic genetic material into a patient's cells. Today, most gene therapy vectors are viruses that have been modified to carry human DNA. As GeneTherapy.net further explains:

> Viruses have evolved a way of encapsulating and delivering their genes to human cells in a pathogenic manner. Scientists have tried to harness this ability by manipulating the viral genome to remove disease-causing genes and insert therapeutic ones. Target cells such as the patient's liver or lung cells are infected with the vector. The vector then unloads its genetic material containing the therapeutic human gene into the target cell.

As we saw in chapter 3, nanotechnologists are already developing a self-assembly technique called protein engineering that can be used to manufacture synthetic viruses. In time, such efforts may provide future gene therapy clinicians with a very sophisticated hardware platform for the fabrication of viral vectors. Yet other innovations in synthetic biology or nanoscale 3D printing may alternatively deliver the tools and techniques required for routine viral vector creation. As we have seen so many times throughout this book, as technologies organically converge toward the nanoscale, so developments in one field of science and engineering have the potential to revolutionize another.

While the use of viral vectors to manipulate human DNA may sound straight-forward, once again it is not. Even when the technology to create such genetic carriers has been mastered, there will still be the risk of the virus infecting and manipulating not just the target patient cells, but additionally other cell types. This has in fact already happened in early clinical trials to genetically treat patients with a rare disease called X-linked severe combined immunodeficiency (X-SCID). This debilitating condition is caused by a single mutated gene, which doctors at Great Ormond Street Hospital in London attempted to treat using a viral vector. Their 2001 trial worked, with an 18 month old baby and 19 other children cured of X-SCID. Unfortunately, four of the patients then developed leukemia due to the viral vector introduced into their body genetically altering other cells.

There is an additional risk that viral vectors may infect reproductive cells, in turn causing genetic changes in a patient's future offspring. Viral vectors may potentially also trigger immune reactions or be passed on to humans other than the patient. While they hold much promise as a genetic therapy technology, viral vectors are hence in need of a great deal of research, refinement and development.

Non-viral means of genetic therapy include the injection of a 'naked' DNA plasmid directly into muscle or other tissue, or the use of non-viral vectors such as lipoplexes and polyplexes. The latter are comprised of DNA coated with liposomes (in the case of lipoplexes) or polymers (in the case of polyplexes). The advantage of coating naked DNA with a fatty substance such as a liposome, or some other form of polymer gel, is that it both protects the genetic material in the DNA and encourages it to stick to the outside of target cells. Some of the genetic material contained in the lipoplex or polyplex may then enter the target cells to replace or re-write their defective genes.

In clinical trials, lipoplex non-viral vectors have been used to transfer healthy genes into cancer cells in order to activate 'tumour suppressor control genes' that decrease cancer activity. Lipoplexes containing a healthy copy of a gene called CFTR have also been put into nebulizer inhalers that have been trialled by sufferers of cystic fibrosis. This has allowed some healthy genetic material to be introduced into the patient's lung cells, which in turn has delivered a slight increase in lung function.

In the coming decades, the development of both viral and non-viral vectors holds the potential to allow genetic therapies to be developed for the treatment of a wide range of conditions. Already trials have shown some success in treating not just XSCID, some cancers and cystic fibrosis, but also hemophilia, beta-Thalassemia and Parkinson's disease.

Much promise has also been demonstrated in the use of genetic therapies to treat some forms of degenerative inherited blindness, such as choroideremia and Leber congenital amaurosis. In the latter case, researchers at Moorfields Eye Hospital in London have treated an abnormality in the gene RPE65 using a viral vector. This was used to introduce a healthy copy of RPE65 into the retinas of three patients, all of whom then experienced an improvement in vision and no side effects. The eye is, in fact, an excellent early candidate for gene therapy, as viruses cannot move from the eye to other parts of the body, and the retina is both readily accessible and yet partially protected from the immune system.

THE DESIGNER BABY

While mainstream genetic therapies will probably not become available until the 2030s, post-genomic technology does now allow certain parents to determine some of the characteristics of their offspring. Such technology forms part of the process of in vitro fertilization (IVF), and in the coming decades is likely to become increasingly controversial.

On 25 July 1978, Louise Brown – the first ever 'test tube baby' – was born following the first successful IVF procedure nine months earlier. In IVF, eggs and sperm are introduced to each other in a laboratory. The resultant fertilized eggs (or 'zygotes') are then cultured for a few days in a growth medium, before being transferred to their mother's uterus in the hope that at least one of them will prove viable and grow to full term.

Because IVF fertilizes and cultures multiple eggs in the laboratory, the possibility exists to genetically test them before they are transferred to the uterus. Such 'pre-implantation genetic diagnosis' (PGD) allows only those embryos with the desired genetic characteristics to be implanted back into their mother. Any child born following an IVF conception can therefore have some of their characteristics determined in a laboratory.

Given that about 1.5 per cent of babies in the United States and the United Kingdom are now born via IVF, the potential implications of choosing embryos based on their genetic makeup is very significant. Even today, over 5 million people owe their existence to IVF, which means that the first generation of transhumans has already arrived. Or to express matters another way, the human race has now evolved into a species that regularly reproduces itself in part using a technology other than 'natural' biology.

Today, pre-implantation genetic diagnosis is mainly used to identify embryos that carry cystic fibrosis, sickle cell disease, spinal muscular atrophy, or a wide range of other genetic ailments. Doctors and parents are therefore using the technique to prevent the birth of a child with a severe medical condition or a disability. It is, however, already possible to genetically test an eight-cell embryo to determine its gender, and in some instances even its eye colour and hair colour. This means that, potentially, some parents could choose to abandon embryos with certain external appearance characteristics in favour of those with desired traits.

In the United States, Mexico and India, a clinic called the Fertility Institutes currently allows IVF parents not only to have their embryos screened for over 400 hereditary diseases, but in addition to choose the sex of their child. The organization describes this service as 'family balancing', and advertises itself as the world's leading 'center for virtually 100 per cent guaranteed gender selection'. The Fertility Institutes does possess the technical skill to offer some parents a choice of eye or hair colour. However, at the present time, they have chosen not to market this capability due to religious opposition.

Increasingly, prospective parents – and maybe even governments – are going to take advantage of IVF conception to facilitate the genetic selection of embryos, and hence to make conscious changes to the human gene pool. Regardless of the ethics, this has to signal that the human race has started to take technological control of its own biological inheritance. Imagine, for example, what would happen if all parents suddenly chose to have a girl. Removing the randomness of nature from the creation of the next generation really could have very far-reaching consequences.

The human race used to be regulated by the survival of the fittest. If this continues to hold true, then by 2030 'fittest' is likely to be a measure not only of a person's ability to attract a mate with the most survival-centric characteristics, but additionally of those technological resources they can afford to introduce into the conception of their child. In time, IVF could even become the conception method of choice for rich couples who are intent on determining the characteristics of their offspring.

GENOMIC UPGRADING

As I noted back in chapter 2, since the 1970s companies including Genentech and Monsanto have been using genetic modification (GM) to alter micro-organisms and plants on a commercial basis. Such GM lifeforms have been created to produce medicines, as well as to improve crop yields or other agricultural product characteristics. GM animals, such as the AquaAdvantage salmon, have even been created if not yet eaten. The future potential therefore has to exist to use post-genomic medicine not just to cure disease, but to otherwise alter and even 'upgrade' the current human form.

As we learn to routinely read and reprogram the human genome, so the whole question of what is 'normal' and what is a medical condition or 'disability' that ought to be 'treated' is going to become extremely contentious. This will particularly be the case when genetic therapies become available at the germline level. Whereas 'somatic' genetic therapies only ever affect the patient being treated, germline therapies make genetic changes that are passed on to future generations. Today the decision to undergo a medical treatment only has personal consequences. But in the future this will not always be the case.

Few are likely to object if somebody opts to have a germline therapy that will prevent both themselves and their descendants from developing a particular form of heart disease or cancer. But what if a cosmetic germline therapy becomes available for altering a person's hair colour? An individual may argue for their right to undergo this 'treatment' to save them the time and money they choose to expend constantly dyeing their hair. But it could also be argued that nobody has the right to tamper with the human gene pool in a manner that will inflict some others in the future with hair that is 'naturally' blue, green or pink.

More radically, and far more controversially, future gene therapies may allow 'patients' to significantly improve their physical or even mental abilities. Athletes, for example, could be 'treated' with a synthetic virus that would re-write their DNA to deliver a performance enhancement. Even today, a synthetic virus called Repoxygen has been created that inserts a gene to stimulate the production of erythropoietin. This hormone then acts on the bone marrow to increase the production of red blood cells, which in turn increases the level of oxygen carried in the blood and delivered to the muscles.

Repoxygen was developed as a potential genetic treatment for anemia. Nevertheless, athletes given Repoxygen would experience improved strength and stamina on a permanent basis. At the time of writing, the practice of 'gene doping' is thought to remain theoretical. It is, however, likely to be very much on the radar of those who regulate sporting activities in the future. Many transhumanists would also argue that athletes – and everybody else – ought to have the option to use all forms of technology to enhance their bodies to be 'better than well'.

Future human enhancements may even involve genetic material crossing the species barrier. Such a concept is also far from science fiction, with the 'transgenic' transfer of genes from one plant or animal into the DNA of another dating back to the 1970s. Firefly genes, for example, have been cut-and-pasted into several other plants and animals in order to make them glow. For over 20 years, an Australian company called Ozgene has even been selling 'humanized' mice that include a little human DNA in order to make them suitable for testing certain drugs.

To date, there is no apparent record of the introduction of non-human genes into human DNA. But the creation of gene therapy vectors that could allow this to happen has to be a possibility. Indeed, as I argued in *25 Things You Need to Know About the Future*:

> Cheetahs can run at 70 miles an hour and are the fastest animal on the planet. So why not isolate the relevant cheetah genes and introduce them into humans to enable them to run faster? Or why not mash a few choice genes from some other animals to give future humans better night vision, glowing fingernails or even wings? These suggestions may sound preposterous. But there has to remain the possibility that at least one mad scientist will choose to experiment.

Already practitioners of synthetic biology have created transgenic micro-organisms, plants and animals, including 'spider goats' that produce a harvestable spider silk protein in their milk. A decade or two from now, farming such 'factories on four legs' in order to locally produce a wide range of specialist chemicals may be routine. Yet so too may be obtaining chemical products from 'factories on two legs' – or, in other words, from human beings with radically-hacked DNA.

Until this point in history, the only product of a human factory on two legs has been another human being, milk, blood, bone marrow, or a donated organ. But some transhumans may be genetically modified for the occupation of producing speciality chemicals in their milk, urine or other body fluids. In fact, some future transhumans may even secrete precious chemicals from entirely new synthetic organs grown or grafted onto their body for that purpose.

Some transhumans may even be genetically modified in order to manufacture products such as organic computing components. As I mentioned in chapter 2, if some future computers or robots have bioelectronic brains, it may prove easiest to manufacture them inside host animals. And those animals could be modified versions of ourselves.

Human beings already grow the best computing wetware on the planet. It is therefore not inconceivable that some of us may grow brains and other components for future computers or robots somewhere inside or on our bodies. Beyond 2050, pregnancy could indeed become a very lucrative occupation. Alternatively – and perhaps more likely – it is possible that future organic humanoid robots will cultivate both computing hardware and replacement human body parts.

Already pigs are being transgenically augmented in the hope that they will one day be able to grow hearts for transplant into human patients. A few decades from now, if you are in need of a new organ, you may therefore have the choice of a personalized, bioprinted body part, a humanized organ grown in a 'natural' animal, fresh tissue cultivated by an entirely synthetic biological being or transhuman, or a 'second hand' organ donated by another human being as is the only option today. This complex prediction extrapolates out a very long way from current scientific and engineering knowledge, let alone medical practice. Yet it hopefully serves to highlight the incredible potential that will arise as post-genomic medicine enters the mainstream.

BEYOND NORMALITY?

Of all the Next Big Things discussed in this book, post-genomic medicine is likely to prove the most contentious. While some people may harbour fears related to the development of nanotechnology or AI, human genetic modification is a far greater concern for a far larger proportion of the population. Not least many people are members of religions that caution against 'playing at god'. In fact, across Europe, even the cultivation of GM crops remains very highly restricted. Those companies intent on developing the most radical post-genomic medical products and practices are therefore likely to face a very hard sell even if there is a strong minority demand for their transhumanist wares.

The range of ethical scenarios to be raised in our post-genomic future is going to be extremely broad. For example, many people – and not just athletes – currently baulk at the idea of human genetic enhancement. But what if a future, germline therapy was able to upgrade our digestive system so that it could extract more nutrition from certain foods? Or to limit the level of fat or sugar that entered the bloodstream? Such a gene therapy could improve the lives of future billions and their offspring. So would it be justifiable to bring this kind of GM human upgrade to market? Or even for a government to freely spread it across an entire population using an airborne synthetic viral vector sprayed from a drone?

Before we step in and say a massive collective 'no' to genetically upgrading all or part of the human species, it will be important to consider the very long-term implications. Clearly, making ill-informed changes to the majority human genome could turn out to be disastrous. Yet equally damaging could be a failure to recognize the need to consciously re-write human DNA as an essential next step in our evolution. Constant genetic change is not just a natural process, but a prerequisite for maintaining the strength of any species. Granted, we could continue to rely on the random mixing of DNA that inevitably occurs in a large, sexually active population. Though as we accelerate toward the Singularity, I suspect that our genetic progress will require a far more conscious push forward and for a variety of reasons.

For a start, I believe that at least some proportion of humanity will require a genetic upgrade in order to facilitate our evolution into space in search of fresh resources. Today, goldfish, cockroaches and fruit flies have a significantly greater resistance to extraterrestrial radiation than human beings, with the fruit fly being at least 60 times more radioresistant than a typical person. Some level of genetic therapy to make future transhumans a little more resistant to extraterrestrial radiation would therefore seem wise. Genetic enhancements for future space transhumans may also include alterations to muscle composition to lessen the rate of atrophy in zero gravity, and perhaps the ability to easily enter and exit a semi-comatose state during lengthy journeys between celestial bodies.

To survive most comfortably in space, future off-world transhumans may also require cybernetic upgrades. In turn, this could necessitate some level of genetic reprogramming in order to facilitate an optimal human-machine interface. We already know that complex inorganic interfaces to the nervous system or brain tend to be poorly tolerated by the human body. Genetic changes to prevent the rejection of spliced-in cybernetic components may therefore be sought by some transhumans, and not just by those who choose to leave the planet.

In parallel with our transhuman evolution, in the coming decades Mother Earth will become home to many new species of synthetic citizen. Many people, let alone many transhumans, are likely to want to compete successfully with these humanoid robots and other AIs, and so in turn may demand cybernetic upgrades that will be most readily facilitated via a parallel genetic re-engineering. Cyborgs may indeed turn out to be the greatest leaders, entrepreneurs and future shapers that this century will raise. In our final chapter we will therefore delve into the future possibilities for human-machine synthesis.

10
CYBORG SYNTHESIS

The human species has always longed to be unique. Highlighting this fact, in 1993 Bruce Mazlish wrote an incredible book called *The Fourth Discontinuity*. Within, he demonstrated how, at various points in history, humanity has had to relinquish its hold on some form of specialness, and how this has had unsettling implications.

Specifically, Mazlish outlines three occasions on which perceived binary divides or 'discontinuities' between human beings and the rest of creation have been bridged. The so-termed 'First Discontinuity' started to be challenged when Nicolaus Copernicus proposed his theory of heliocentricity, with the Earth rotating around the Sun rather than the other way around. The 'Second Discontinuity' then began to fall when Charles Darwin popularised his theory of evolution, so linking humans to every other creature on the Earth. A 'Third Discontinuity' was next called into question when Sigmund Freud queried any absolute divide between our conscious and subconscious selves, and in the process highlighted human beings as psychological as well as physiological creatures.

When Copernicus dispelled the myth that humanity was at the centre of the Solar System, Darwin demonstrated that there is no divide between human beings and the animal kingdom, and Freud linked our conscious and subconscious, so *Homo sapiens* had to accept a drop in status. Today, as the title of Mazlish's book highlights, we are just starting to pass through the 'Fourth Discontinuity', which is where we will need to accept that there is no absolute distinction between human beings and machines.

Mazlish suggests that there are two reasons to believe that the Fourth Discontinuity is starting to be crossed. Firstly, he states that it is no longer realistic to think of humans without machines. And secondly, he suggests that the same concepts now explain and govern the function of both human beings and artificial technologies. To my mind at least, both of these propositions also seem to be eminently reasonable.

For a start, in developed nations, most people's lives are now sustained by complex technological infrastructures. As we have seen throughout this book, future developments in 3D printing, synthetic biology and nanotechnology are also going to be used for medical purposes as well as product fabrication. Mazlish therefore has to be right in his contention that the function and maintenance of human beings and machines is starting to become similar both conceptually and on a practical, operational level.

CYBORGS EMERGE

Any creature that is an amalgamation of 'natural' and 'artificial' parts is technically known as a cyborg. What Bruce Mazlish outlines in *The Fourth Discontinuity* is therefore how the human race is evolving into a cyborg species. Thus far, very few individuals have been permanently fitted with artificial parts aside from dumb, medical prosthesis. But many people are already reluctant to be separated from their technological devices, with smartphones in particular surgically clasped in many a daily grip.

As technology is interfaced with the human body in an increasingly sophisticated manner, so the Fourth Discontinuity will well and truly be crossed. In fact, even when humanoid robots start to regularly walk the Earth in the early 2030s, it will become increasingly difficult to cleave any absolute, binary divide between humanity and artificial technology.

The above will especially be the case when we start to build humanoid robots out of synthetic biological components. By 2050, humanoid robots could be flesh-and-bone bipeds that are as intelligent as we are, and with whom we may be able to exchange body parts. Justifying an absolute distinction between human beings and such biological robots will therefore be impossible.

This final chapter will explore a wide spectrum of possibilities for the convergence of 'natural' human biology with 'artificial' technology. Along the way we will consider how the Internet has already turned us all into the cells of a global, cyborg entity, and how future transhuman space pioneers may adopt modular bodies in order to facilitate a myriad of off-world pursuits. But before that, we will start by looking at the shorter-term possibilities for individual cybernetic maintenance and transhuman upgrading.

FROM DUMB TO ELECTRONIC PROSTHESIS

For hundreds of years doctors have been using prosthetics to repair or support compromised human biology. The majority of these devices – including spectacles, hearing aids and false limbs – have been worn rather than permanently fitted, and remain removable to this day. This said, dental crowns and other non-removable prosthetics are believed to date back several thousand years.

Today it is quite common for patients to be fitted with permanent bodily additions that include not just teeth, but also replacement hip and knee joints, heart valves and hernia meshes. Until quite recently most of these have been off-the-shelf devices available in a limited range of sizes. But with the development of 3D printing, some patients are now being given personalized plastic or metal prosthesis that are fabricated-on-demand according to measurements derived from a 3D scan.

In July 2015, a team from the Southern Medical University in Guangzhou, China, announced that they had developed a method for 3D printing replacement bones using layers of a real bone powder selectively sprayed with adhesive. Already this work is entering an animal testing stage, with human trials of 3D printed bones expected in the early 2020s. As we saw in chapter 1, by the end of the next decade it is quite likely that doctors will begin bioprinting a wide range of human body parts for transplant purposes. Replacement organs grown in animals that have been genetically 'humanized' with patient DNA are also a possibility before 2030.

While most prosthetics remain inanimate objects or very simple mechanical mechanisms, increasingly our bodily additions may feature some kind of electrical or electronic circuitry. The first such bioelectric devices were pacemakers to regulate a patient's heartbeat, which started to be fitted in the late 1950s. By the 1960s, researchers had also begun to successfully artificially stimulate the human auditory nerves. The first cochlear implants were subsequently given to deaf volunteers in the early 1970s.

Today over 200,000 people have been fitted with a cochlear implant. The procedure involves the surgical insertion of an electrode array into the patient's cochlear to deliver electrical impulses to their auditory nerves. The array is connected to a receiver-stimulator device that is also surgically inserted, and which receives signals wirelessly from a small, external sound processor that is worn by the patient. Recently, external wireless streamers have become available that send the audio from a computer, television or other electronic device directly to a patient's sound processor.

While cochlear implants are an established cybernetic technology, bionic eyes are in a far earlier stage of development. Progress is, however, being made, with a Californian company called Second Sight now marketing its Argus II Retinal Prosthesis System. This features an electrode array that is attached to the retina inside the eye, as well as an antenna and electronics that are surgically positioned on the outside of the eyeball. Special glasses and a visual processing unit are then worn externally.

In a healthy eye, photoreceptors in the retina convert light into electrochemical impulses that are sent through the optic nerve to the brain. But in people who have the condition retinitis pigmentosa, the photoreceptors no longer function correctly. The Argus II aims to return some level of vision to people with this or similar conditions, with its technology able to bypass the retina's damaged photoreceptors. As Second Sight further explain:

> A miniature video camera housed in the patient's glasses captures a scene. The video is sent to a small patient-worn computer ... where it is processed and transformed into instructions that are sent back to the glasses via a cable. These instructions are transmitted wirelessly to an antenna in the implant. The signals are then sent to the electrode array, which emits small pulses of electricity. These pulses are intended to bypass the damaged photoreceptors and stimulate the retina's remaining cells, which transmit the visual information along the optic nerve to the brain. This process is intended to create the perception of patterns of light which patients can learn to interpret as visual patterns.

The current Argus II includes an array of just 60 electrodes, which is a very small number compared to the retina's 100 million or more photoreceptors. Yet even the 60 electrode array allows those fitted with the implant to make out shapes, and sometimes even large letters. An Argus III implant with 200 electrodes, and a fourth generation model with 1,024 electrodes, are also planned. Given that in the United States alone over 100,000 people have retinitis pigmentosa, this has to be great news.

The Argus II was only approved for implant into patients in 2011 in Europe and 2013 in the United States, so these are still very early days. This said, it is not difficult to imagine implants with tens of thousands of electrodes being available by the late 2020s or early 2030s. An implant with about 76,000 electrodes would, for example, provide a resolution of 320x240 pixels – which for several years was the standard for YouTube and other online video.

DIRECT NEURAL CONNECTIONS

In addition to making auditory and optic connections, scientists have begun to successfully engineer prosthetic limbs that can be directly controlled by a patient's nervous system. Most often, a technology called 'myoelectrics' is employed, which places electrodes on the surface of a person's skin. These then detect electromyograph (MPG) nerve and muscle signals and use them to activate servo motors in a prosthetic. In practice this means that a hand amputee can learn to flex muscles in the remaining section of their forearm in order to trigger a prosthetic hand to open or close.

Several companies, including Open Bionics and Touch Bionics, now make myoelectric prosthetic hands that are intended to help the several million hand amputees worldwide. The hardware made by Open Bionics is low-cost and 3D printed to patient specification. Figure 10.1 shows an early but fully-functional version of their prosthetic that I saw in operation in May 2015.

Touch Bionics make several versions of a highly sophisticated prosthetic hand called the i.limb. This is controlled by myoelectric electrodes placed on two muscle sites, which can

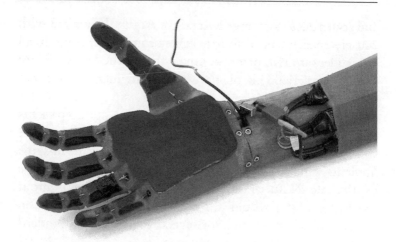

Figure 10.1: An Open Bionics 3D Printed Prosthetic Hand.

be used to trigger the hand to move into one of four different positions such as a 'lateral grip' or a 'precision pinch'. Finger positions can also be selected via a smartphone app. In addition, some of the Touch Bionics hands feature a gyroscopic gesture control system called 'i-mo'. This allows the user to change the position of their prosthetic fingers with a simple movement of their arm.

To address the problem of trying to control the operation of a prosthetic from relatively few remaining muscles, the Center for Bionic Medicine at the Rehabilitation Institute in Chicago has developed a technique called 'targeted muscle reinnervation' (TMR). Here, nerves from amputated limbs are transferred – or 'rewired' – to another part of the body. So, for example, surgeons may re-route nerves from an amputee's shoulder to the ulnar, musculocutaneous, median and radial nerves on the left or right of their chest. Electrodes are then placed on the chest in the appropriate places, and in time the patient can learn to use their retargeted chest nerves to control a myoelectric prosthetic limb.

Back in 2007, a former US marine called Claudia Mitchell was given an entire, TMR-controlled artificial arm and hand after losing a limb in a motorcycle accident. She has since learnt to control this highly advanced prosthetic, and has even begun to experience 'sensory reinnervation'. This means that touch sensations on Claudia's left chest register as if coming from her amputated limb. In time it is hoped to use touch sensors in Claudia's artificial hand to transmit signals back to her re-routed nerves. In theory, this may allow her to obtain touch and temperature feedback from a cybernetic prosthesis.

Even when used in conjunction with TMR surgery, myoelectric technology faces inevitable limitations due to its reliance on intercepting nerve signals from electrodes placed on the skin. In the future, more sophisticated prosthetics – and possibly other human-technology interfaces – are therefore likely to rely on microchip arrays surgically inserted into the body. Working toward this goal, already the Alfred E. Mann Foundation for Biomedical Engineering at the University of Southern California has developed 'implantable myoelectric sensor' (IMES) devices. A few of these have even been surgically inserted into human test subjects, some of whom have managed to achieve direct nerve-interface control of either robotic hands or robotic legs.

Also working on technologies to establish direct neural connections are Battelle, the world's largest nonprofit research and development organization. Over in the United States, the company has developed a microchip technology called Neurobridge, which is initially intended to help patients with spinal cord damage by providing an alternative interface between their brain and limbs. Battelle's 'NeuroLife Neural Bypass Technology' consists of a tiny chip for brain implantation, together with a 'high-definition electrode stimulation sleeve' that is worn on the appropriate limb.

In April 2014, neuroscientists Ali Rezai and Jerry Mysiw from the Ohio State University Wexner Medical Center implanted a Neurobridge chip into a human patient. Their test subject was Ian Burkhart, a 23 year old quadriplegic who had been paralyzed following a driving accident. As the Battelle website explains:

> During a three-hour surgery on April 22, Rezai implanted a chip smaller than a pea onto the motor cortex of Burkhart's brain. The tiny chip interprets brain signals and sends them to a computer, which recodes and sends them to the high-definition electrode stimulation sleeve that stimulates the proper muscles to execute his desired movements.

Two months after the operation, Ian managed to make a fist with his right hand using only his brain waves, transferred through his brain implant, to make his fingers move. This was the first time that a patient had moved a paralyzed limb 'just by thinking about it', and demonstrated the extraordinary potential to read signals from the brain. For patients with a damaged spinal cord that prevents nerve signals reaching otherwise healthy limbs, the medical opportunities that may result could be life changing. Perhaps even more significantly, the creation of Battelle's technology could prove a stepping-stone in the development of the ultimate human-computer interface.

BRAIN INTERFACES

In addition to Battelle, for many years a range of scientists have been working to develop brain-computer interfaces (BCIs). If such hardware could be perfected, then in the future people could be surgically altered (or genetically modified) to allow video and audio signals to be fed directly into their brain. Additional memory, computational power, translator devices and GPS circuitry could also potentially be retrofitted to the human body. Such pure cyborg technology lies decades into the future. But the possibilities it may offer would be quite extraordinary.

No longer would we need to stare at a screen and use headphones or speakers to experience digital entertainment or communications. Instead, we could develop mental techniques that would enable us to turn our perception to any external audio-visual data feed of our choosing. Meanwhile, hardware grafted to our brains could provide us with all of the knowledge of the Internet, the ability to understand and speak any language, and the capability to operate cybernetic body parts. Anybody with a suitable brain-computer interface could also, in theory, access the thoughts and control the body of another willing individual with appropriate cybernetic hardware. Quite amazingly, the first experiments to actually allow this to happen have now even taken place.

In November 2014, researchers from the University of Washington demonstrated how volunteers could be linked together so that signals from the brain of one person could control the hand motions of another. In the experiment, three different 'sender' test subjects donned electroencephalography (EEG) sensor headwear to read their brainwave patterns. Surrogates of these signals were then transmitted over the Internet and fed to swim-cap-style headwear worn by 'receiver' subjects. This receiver headwear featured 'transcranial magnetic stimulation' (TMS) coils positioned near the part of the brain that controls hand movements.

Each sender subject was seated before a computer game in which they had to defend a city by firing a cannon to intercept incoming rockets. But the touchpad to fire the canon was below the hand of a receiver subject who could not see the game. The only way for a sender subject to successfully fire a rocket was therefore for them to think about moving their hand, and for this thought to be read by their EEG headwear, transmitted over the Internet, and fed into the brain of the receiver subject whose hand would then move to fire the cannon. Amazingly, the brain-to-Internet-to-brain connection worked for all three pairs of test subjects, although the accuracy level achieved in shooting down missiles varied between 25 and 83 per cent.

In another experiment, reported in September 2015, University of Washington researchers next used a similar, non-invasive brain-to-brain interface (BBI) to allow test subjects to play an interactive game similar to 20 Questions. As in the previous experiment, EEG hardware was used to read the brainwave patterns of 'respondents'. These were then transmitted to TMS coils worn by 'inquirer' subjects which, on this occasion, caused them to experience visual stimuli.

The experiment was conducted using five pairs of inquirers and respondents, each of whom played a game where the respondent thought about one of eight animals which their enquirer then had to guess by asking three yes/no questions. The answers to these questions were communicated by the brain-to-brain interface, with the respondents staring at one of two flashing LEDs in order to transmit their answer. These two LEDs flashed at different rates, so triggering different brainwave patterns that were picked up by the respondent's EEG skullcap. The inquirers were then sent a signal based on the respondent's brainwave patterns, and 'saw' a visual stimuli if their answer was 'yes', and nothing if it was 'no'.

On 95 per cent of occasions, inquirers were able to accurately receive their answers via brain-to-brain communication. Granted, respondents had to stare at the appropriately flashing LED for up to 20 seconds for their EEG decoder to read their answer. Inquirers also required several hours of training to be able to reliably detect their responses. Even so, the experiment did demonstrate the potential for direct, private brain-to-brain communication.

EEG brain interfaces are likely to continue to improve in the years and decades ahead, with the technology ripe for development by open-source enthusiasts and hackers. Indeed, in June 2015, a Chinese postgraduate student demonstrated the use of an EEG brain interface to control the direction of travel of a cockroach. Here the insect was fitted with an electronic 'backpack' that received wireless signals. These in turn stimulated the creature's circi appendages, so causing it to move in the direction the student was thinking about.

Also pioneering the development of brain interfaces is a long-standing consortium called BrainGate, based at Brown University. Like Battelle, for many years this research team has been developing an implantable chip, with the first 'BrainGate Neural Interface System' created as long ago as 2002. In the long-term, the goal of the BrainGate team is to 'create a system that, quite literally, turns thought into action', and which will be useful for 'people with neurologic disease or injury, or limb loss'.

The BrainGate system consists of a sensor, a decoder, and some kind of external hardware that the sensor and decoder will allow a patient to control. Currently the sensor is a 4 x 4 millimetre array of 100 electrodes that, when surgically implanted, can read 'neuronal action potentials' and 'local field potentials' from the brain.

The BrainGate decoder is a set of computers and embedded software that turn the brain signals from the sensor into a useful command for an external device. Such hardware could potentially be a personal computer, a powered wheelchair, a prosthetic or robotic limb, or an electrical stimulation device that may move paralyzed limbs.

In 2004 a research team from the University of Pittsburgh implanted a BrainGate sensor into the motor cortex of a monkey. The decoder was then hooked-up to a robotic arm, with food placed at different locations in front of the test subject. With its own arms constrained, the primate actually learnt to use the robot arm to pick up the food and feed itself.

In 2008 a company called Cyberkinetics – which for a time developed the BrainGate system – implanted one of its tiny electrode arrays into a human being. The subject was a 25-year-old quadriplegic, who managed to use the BrainGate system to turn on lights, to change the channel and volume level on a television, and to control the display of e-mail messages.

In May 2012, scientists at Brown University went a stage further when they used the BrainGate system to interface the motor cortex of two human subjects to a robotic arm. In the trial, a paralyzed 58-year-old woman and a paralyzed 66-year-old man both managed to grasp target items with their cybernetic appendage. The 58-year-old woman was even able to pick up a bottle from a table, raise it to her mouth, take a drink, and gently put the bottle back down again.

For many years a major limitation of the BrainGate system was that the decoder consisted of a large rack of signal processing computers and other electronics. Given that this had to be directly connected to the implanted sensor via a bulky cable attached to a socket in the skull, it was impractical to use the BrainGate system outside of a lab. Fortunately, by January 2015, researchers from Brown University had worked with a company called Blackrock Microsystems to commercialize a tiny, wireless, battery-operated decoder called the Cereplex-W. This is worn on the skull, where it connects directly to the sensor implant. It then interfaces wirelessly with a receiver that may potentially control external devices. At the time of writing, the CerePlex W brain-interface is still very much a research tool, if one with a significant potential to help scientists push forward the development of direct electronic interfaces to the human body.

AUGMENTATION HORIZONS

The brain interface developments outlined in the previous section are clearly extraordinary. To me at least, they also suggest that a new age of cyborg synthesis will sooner or later dawn. Though given the developments we have explored elsewhere in this book, I suspect that most of tomorrow's human-machine interfaces will not depend on integrating the carbon-based hardware of the natural human body with silicon-based electronics. Highly refined EEG headsets may perhaps form the basis of some future brain interfaces. But, in the longer term, we are more likely to employ bioprinting, synthetic biology and nanotechnology to create organic rather than inorganic prosthetics and bodily augmentations.

Some combination of post-genomic medicine and cyborg synthesis could one day allow human beings to be fitted with entirely new organs or organ combinations. Already pancreatic bioprinting pioneer Ibrahim Ozbolat has suggested that it may be more appropriate to fit future patients with several small, bioprinted pancreatic organs rather than a single, like-for-like transplant. Such an approach, he supposes, could provide greater redundancy cover and less post-operative risk. Nature may have done a pretty amazing job in determining just what organs are required to build a human being and in what proportion. But this does not mean that further biological optimizations beyond the scope of 'natural' evolution are out of the question.

With the exception of the kidneys and the lungs, pretty much every one of our internal components is mission critical and yet has little or no redundancy cover. Much of the time, the whole system works well for up to a century or more. But just as nobody would construct a hospital, a nuclear power station or a spacecraft without duplicating key components to provide fail-safe operation, so some future transhumans may apply AI biological engineering to improve the fault tolerance of their own bodies.

In part to cope with the fault tolerance issue, our bodies do possess an amazing disease-defence infrastructure called the immune system. Yet in the future, this too may be artificially augmented. Today, doctors do try to assist their patient's immune systems using chemical drugs. Antibiotics, for example, are regularly prescribed to kill bacterial infections, and so help out overworked natural antibodies. But in addition to such short-term medical treatments, some transhumans may have nanobots or 'synthetic antibodies' permanently added to their bloodstreams. These could constantly roam their host's vascular network to monitor health, fight disease, and repair damage at a molecular level. Potentially, some future nanobots may even spend their free time 'upgrading' their host's internal organs.

Figure 10.2 shows a visualization of a future nanobot as it intercepts a damaged blood cell. In practice, nanobots capable of molecular repair and upgrading will almost certainly be entirely organic machines self-assembled by synthetic biologists and AIs. No nanobot is therefore likely to have a gleaming metal body and appendages as depicted in figure 10.2. However, I thought you would feel cheated if you got to the end of this book and had not seen an illustration of a nanobot!

Nanotechnology capable of fulfilling a medical role inside the human body is now being developed. In particular, much work is focused on the creation of microscale and nanoscale drug carriers that can more precisely deliver medicines than traditional tablets or injections. For example, in 2014 at the Department of Nanoengineering at the University of California, a team of researchers used artificial, microscopic machines to deliver a drug to the stomach of a mouse. Their tiny 'nanobot' cylinders were about 20 micrometres long, 5 micrometres in diameter, and propelled by zinc-based micromotors that emitted hydrogen bubbles when exposed to the acid contents of the stomach. This caused them to travel to

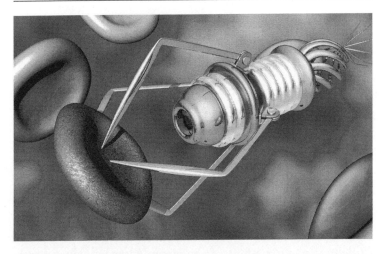

Figure 10.2: Nanobot Concept. Image by Christopher Barnatt.

the stomach lining, where they embedded themselves, dissolved, and so delivered a nanoparticle compound into the tissue of the gut.

Other cutting-edge research is taking place at the Institute for Robotics and Intelligent Systems at ETH Zürich in Switzerland, which is one of the world's leading science and technology universities. Here, for many years a team led by Bradley Nelson has been creating microrobots and nanorobots that can be steered around the body using external magnetic fields. Within a decade it is hoped that these miniature devices will be used for highly-targeted drug delivery, or even to perform delicate surgery within the eye or other organs. Such hopes are still far from the popular science fiction vision of nanobot medical maintenance technology. But they may deliver important, first-step innovations.

DIY EVOLUTION

Back in the *Prologue* I mused how, in a few decades time, you may have a warm bucket in a kitchen cupboard in which you are growing a new arm for your favourite android companion. As you hopefully now appreciate, the development of some combined synthetic biology and nanotech self-assembly process that could allow this to happen is perfectly feasible. When it becomes possible, there is also no reason to assume that 'medical' home-based local digital manufacturing will be limited to the production of fresh android parts.

In 20 or 30 years, clever little synthetic microbes in a household cabinet could be synthesizing fresh components not just for your robot, but also for your own body, or maybe the hardware of your children or your dog. Potentially such additional or replacement parts could be surgically fitted by a domestic humanoid robot that will be at least as skilled as any current human surgeon. Alternatively, we may be able to swallow or inject small or partially-formed body part components that will traverse our digestive or vascular systems, and then amalgamate and fuse into place where their digital programming dictates.

As yet another option, some people in the future may be genetically or surgically modified with easily-accessible 'expansion ports'. Grafted or grown into our torsos or heads, these would provide full access to our nervous and vascular systems, and could allow the fairly regular insertion and removal of hospital or home-grown biomodules that would 'heal' into place to provide a potentially very wide range of services.

Biomodules could, for example, be added to boost memory or stamina, or to provide wireless networking and in turn brain-to-brain communications (telepathy). Alternatively, the old or the sick may have biomodules plugged in that would help them to fight a particular disease. Meanwhile pregnant women (or future men with a womb fitted) may plug in appropriate biomodules to optimize and monitor the health of their unborn child.

Talking of babies, next-generation humans may be synthetically engineered with bioports that would allow

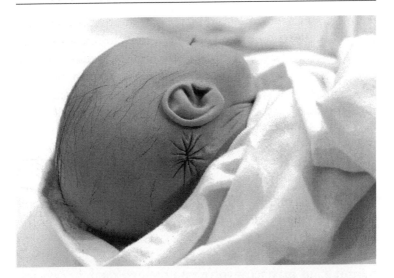

Figure 10.3: Epidermal Myostomy Concept by Agi Haines.
This *Transformations* image depicts a baby with a surgically-created bioport that may allow medicines or biomodules to be inserted.

biomodules to be swapped in-and-out to assist with their physical or mental development. Bioports may also be surgically fashioned that could allow tablets or other drugs to be administered by means other than the mouth or injections. In figure 10.3, artist Agi Haines illustrates what such a new, transhuman orifice may look like.

Out in space, organic and inorganic cyborg synthesis may offer even more extraordinary possibilities. As I discussed in chapter 8, the evolutionary leap from our first planet to the vacuum of space is likely to demand a fairly radical, 'synthetic' evolution of at least some 'human' beings. Future, non-robotic deep space travellers may, for example, be fitted with cybernetic prosthesis that will allow them to survive on a diet of electricity, rather than oxygen, water and organic food.

Tomorrow's space pioneers may even radically rebuild their bodies according to the dictates of their situation. A body suited to sustaining the human brain on a multi-month or multi-year space flight is, after all, likely to be very different to a body optimized for living and working on a lunar or asteroid mining colony. So why not abandon the notion of a single, fixed body and self-evolve as required? For centuries we have learnt to equip and clothe ourselves to survive and thrive in different environments. In the future, it may therefore turn out to be a logical progression to apply the new tools and techniques of local digital manufacturing to change our bodies, and not just our garments, as the situation dictates.

If the above proposition comes to pass, then some future transhumans may self-evolve into the first modular species. Today we all refer to 'parts' of the human body that are mostly difficult to precisely isolate, and which are very difficult indeed to remove or replace. Where, for example, does your forearm *really* begin and end?

Human beings – like all other animals – are actually a continuum of parts; a very-tightly-integrated, single-component composite. In contrast, at present only inorganic technologies are modular, with purely discrete and hence exchangeable components. This is, after all, why the current practice of medicine is so hard. Human parts are not yet separable, replaceable and exchangeable like the components of our computers or our cars. Though as we develop a mastery of bioprinting and molecular self-assembly (and self-disassembly), this will not have to remain the case.

LIFE BEYOND THE FLESH?

Innovations in post-genomic medicine and cybernetics may one day greatly extend the life span and capabilities of the human form. Nevertheless, the wetware of even the best human brain is not going to last forever. Some future transhumans may therefore forsake their birth-wetware in favour of an entirely artificial body – or indeed, no body at all.

You will probably not be surprised to learn that experiments to construct an electronic brain have already started. Most notably, the Blue Brain Project was launched in 2005 by the École Polytechnique Fédérale de Lausanne (EPFL) working with IBM. Now based at EPFL's Biotech Campus in Geneva, this ambitious, long-term undertaking intends to create a simulation of an organic brain within a computer. Or as the project itself explains, 'the goal of the Blue Brain Project is to build biologically detailed digital reconstructions and simulations of the rodent, and ultimately the human brain'.

In October 2015, the Blue Brain Project reported that it had successfully produced the first-draft of a digital reconstruction of the microcircuitry of the somatosensory cortex of a rat. This computer model 'simulates about a third of a cubic millimetre of brain tissue containing about 30,000 neurons connected by nearly 40 million synapses'. The neurological operation of this virtual tissue has even had its accuracy verified via comparison with test data obtained from experiments conducted on a real rat brain.

What the above means is that scientists are now capable of reproducing a fully-functional copy of part of a living brain in a computer. Such an accomplishment is clearly extraordinary, and could have very far reaching implications.

In 2013, The Blue Brain Project became the 'simulation core' of a 10 year undertaking call the Human Brain Project (HBP). This has a mission to obtain 'a multi-level, integrated understanding of brain structure and function through the development and use of information and communication technologies'. The €1.19 billion project is funded by the European Commission, and involves hundreds of researchers from 135 partner institutions working across 26 countries.

The Human Brain Project has the potential to deliver radical advancements in neuroscience, computing and medicine. Not least, it could revolutionize our approach to the development of future AI. For as the project's extensive FAQ explains:

> The challenge in Artificial Intelligence (AI) is to design algorithms that can produce intelligent behaviour and to use them to build intelligent machines. It doesn't matter whether the algorithms are biologically realistic – what matters is that they work. In the HBP, we're doing something completely different. The goal is to build data-driven models that capture what we've learned about the brain experimentally: its deep mechanics (the bottom-up approach) and the basic principles it uses in cognition (the top-down approach). Certainly we will try and translate our results into technology (neuromorphic processors) but, unlike classical AI, we will base the technology on what we actually know about the brain and its circuitry. We will develop brain models with learning rules that are as close as possible to the actual rules used by the brain and couple our models to virtual robots that interact with virtual environments. In other words, our models will learn the same way the brain learns. Our hope is that they will develop the same kind of intelligent behaviour. We know that the brain's strategy works. So we expect that a model based on the same strategy will be much more powerful than anything AI has produced with "artificial" algorithms.

In addition to pushing forward AI, as and when a complete, working copy of a human brain is simulated in a computer, so the frontiers of transhuman evolution are likely to rapidly advance. Few seem to be asking if a virtual model of a human brain could develop consciousness, or one day control a human or android body. But, in the very long term, such things have to be a possibility.

There is, of course, a massive difference between re-creating a digital copy of the hardware of a brain, and 'loading' it with the memories, knowledge and consciousness of a person. Right now, nobody has any idea how the functional contents of a human brain could be 'uploaded' into virtual hardware. But, once we have managed to build an 'empty' brain in a computer, so transferring a person from their native biological hardware to an artificial alternative is something to which at least some future AIs or transhuman devotees may understandably turn their attention. In May 2015, neuroscientist Hannah Critchlow of Cambridge Neuroscience publicly stated that 'it is definitely a possibility' that 'people could probably live inside a machine'.

If a person could upload themselves into a computer simulation of the brain, they may additionally be able to download back into fresh organic or inorganic hardware. They could even potentially traverse deep space in a digital format in order to occupy a suitable body waiting on another planet or space station. Future transhuman astronauts may therefore end up travelling between worlds via the Internet and radio wave.

The idea of uploading and downloading inevitably raises all kinds of tricky questions. For a start, just what would happen to the 'copy' of a person that may continue to reside in their original human wetware? In the future, transhumans may not only be able to live forever, but to endlessly reproduce themselves via copy-and-paste.

When residing in a digital brain, people could also potentially run at different speeds according to the capabilities of available hardware. By gaining access to enough computing power, an uploaded individual (or temporary clan thereof) could therefore complete a week's mental activities in an hour or a minute or a second. While this could allow transhumans to 'keep up with the robots', it also implies that native-flesh humans would be unable to compete with transhuman uploaders. The implications for our society would hence be monumental. And yet, as we approach the Singularity, these are exactly the kinds of issues that our descendants may have to confront.

HUMANITY 2.0

In 1993 Gregory Stock published an amazing book called *Metaman*. A two-paragraph blurb on the back cover summarized his work perfectly, and is reproduced in its entirety below:

> Imagine looking down from the Moon at the night side of the Earth, pitch dark and invisible except for a brightly lit network of human constructions – luminous cities, highways, canals, telephone and power lines. A faint, speckled web of light would seem to float in space. Some regions of this lacework would form intricate geometric patterns, others would seem random and disconnected. Far from inert, this distant pattern of light would change and grow over the decades as you kept watch, shimmering fibres forming, extending, and joining in an almost vegetative fashion.
>
> This resemblance to life is not mere happenstance; the thin planetary patina of humanity and its creations is actually a living entity. It is a "superorganism", that is, a community of organisms so fully tied together as to be a single living being itself. Instead of referring to this entity using a term

already filled with associations, let's start afresh and simply call it "Metaman".

What Stock highlighted with his identification of Metaman was the emergence of the Earth's first intelligent cyborg. Back in 1993, Stock did not equate this planet-spanning entity with the then-pubescent creature of the Internet. Yet today, it is clear that we can look at the human race as either seven billion individuals, or as the seven billion neurons that crowdsource their collective, biological abilities via the inorganic infrastructure of our planetary computer network.

Some people may have read this chapter and wondered if they really wanted to become a cyborg. Well, my final news is that we are all already part of a cyborg, and a rather amazing one at that. Throughout this book, I have made repeated reference to new forms of beyond human intelligence that will take us on the journey to the Singularity. And as Stock highlighted back in 1993, on a planetary scale such a new form of collective intelligence has already started to arrive.

Some futurists now use the label 'Humanity 2.0' to refer to the transhuman species that is emerging as we start to merge technology with ourselves, and merge ourselves with technology. Personally I am a little nervous about the term 'Humanity 2.0', as it infers that we are heading toward a binary divide between the humans of today and the humans of tomorrow. This is, I think, unlikely, with our conscious evolution destined to be a gradual process that will result in many different branches of a post-human race. Though given that we started this chapter by considering the blurring of any binary divide between 'technology' and 'humanity', I guess it may be reasonable to define 'Humanity 2.0' as the human species that will exist when we have accepted the crossing of Bruce Mazlish's Fourth Discontinuity.

As and when we come to accept that we are no more and no less than an amazing form of technology, so a great many things will start to change. Back in Part I of this book, I detailed the future rise of local digital manufacturing (LDM) as 3D printing, synthetic biology and nanotechnology converge into the single, new technology of microfabrication. Future microfabricators will allow almost anything to be fabricated according to the dictates of a digital computer model, and such things will certainly include artificially fabricated parts of ourselves.

As we have seen in this chapter, in the future we may even learn to reproduce the human brain in a computer. In the process we would also totally blur the boundaries between ourselves and artificial intelligence. As I detailed back in Part II, tomorrow's 'synthetic citizens' will not possess 'human' intelligence. But they will possess increasingly high levels of alternative mental competencies. We are also very likely to rely on such artificial cognitive abilities to help build the microfabricators and other technologies that will enable future 'human' civilizations to survive and thrive.

Nowhere will robots and other synthetic citizens prove more valuable than in our forthcoming conquest of space. As we saw in Part III, in the second half of this century it is perfectly possible that we will start to obtain both energy and raw material supplies from beyond our first planet. Such new natural resources will also prove essential if humanity is not to face an inevitable decline due to resource depletion here on Planet Earth.

More than most people I preach and live the growing meme of 'sustainability' by attempting to consume things less and to value things more. Even so, 50 or less years out, I strongly believe that the increasingly popular mantras of *recycling* and *consuming less* will have to be augmented by the physical necessity of *finding more*. When this finally happens, AIs, robots and microfabricators are destined to rise to the fore as those critical tools and allies that will help us to evolve beyond the Earth.

As we tentatively venture away from our motherworld, so at least some of our bodies will have to change. *Homo Sapiens* have evolved to live on warm lands with a strong gravitational hold and a biologically-friendly atmosphere, and not in the cold, non-matter-hugging vacuum of space. The boldest future transhumans will therefore take steps to augment themselves – either biologically or inorganically – in order to become the most successful space pioneers.

As we self-evolve in such a manner, Humanity 2.0 may well transition to become a modular species that takes on different physical forms as different environments dictate. In the future, our descendants may therefore walk the Earth in a tight amalgam of organic parts, but travel the stars in a loose collection of nanocomposite, inorganic hardware. This would make evolution a constant, short-term process, and on occasion even a real-time individual choice.

As Gregory Stock reminds us, as we weave the tapestry of the future with so many incredible but interlinked technologies, so tomorrow's humans and transhumans are likely to plug-in and plug-out of collective super-organisms at an increasing rate. Today we can already see the line between 'individual' and 'collective' humanity starting to blur. To alien outsiders, our future civilization may hence appear to be a massive Lego set, with its pieces sometimes assembled in small clumps known as individuals, while on other occasions pieced together into a single structure capable of achieving greater things. Humanity 2.0 will be not one of us but all of us, and that means all of us including all of our technologies.

Back at the very start of the *Prologue* I pronounced that the Internet Revolution is over, and that it is time to move on to embrace the Next Big Thing. Hopefully, as a reader of this book, you now understand why I made this assertion and how true it must be. The Internet will continue to help us to interconnect and to build an amazing future. But it is now time for humanity to shift its collective focus away from cyberspace and toward the dawning revolutions in microfabrication, AI, space travel and transhuman upgrading.

Twenty years ago, the extraordinary revolution of the Internet demonstrated that the impossible can become possible. And on multiple frontiers, it is now time for the impossible to become possible again.

EPILOGUE

TOWARD THE SINGULARITY

The other week I was flipping TV channels when I came across an early 1980s cop show. As I joined the episode, our hero was out in the countryside, where he suddenly deduced that a gun-wielding maniac was headed to his girlfriend's house. So how could he let her know of the impending danger? Well, the only solution involved running across several fields, clambering into a classic sports car, and driving at speed back into town. Fortunately this strategy proved successful, and the good copper arrived just in time to apprehend the murderous villain before the credits rolled.

Today, many young people watching the above TV show would be somewhat perplexed. Why, they would ask, did our hero not just get out his smartphone and give his girlfriend a call? Or send her a warning text? Or even a 'you are in dire peril' tweet? The answer, of course, is that smartphones – and cell phones more generally – did not exist when the TV show was made. Nor was any member of the general public on the Internet. In fact, in the early 1980s, even personal computers and video recorders were a novelty, while notebooks were all made out of paper, and tablets required a medical prescription.

Often in my work as a futurist I am told that tomorrow will not be that different from today. To the extent that, centuries hence, people will still be eating, sleeping, loving and arguing, I would agree with this proposition. But, as looking back to 1980s TV shows ought to remind us, over a period of decades it is perfectly possible for the application of new technologies to transform at least some aspects of our lives. Despite the Dot Com boom and bust, those in the early-to-mid 1990s who predicted that the Internet and mobile technologies were the Next Big Thing were clearly correct. And as I indicated way back in the *Prologue*, with the publication of my 1995 book *Cyber Business*, I was very much one of those early Internet and mobile technology pundits.

Today the Internet is the Last Big Thing, with the Next Big Thing likely to involve those technologies and related undertakings we have explored in the last ten chapters. This does not, however, necessarily mean that the mainstream rollout of technologies like 3D printing, synthetic biology, nanotechnology and AI will be enough to automatically drive radical change. It all depends on whether these technologies turn out to be 'deterministic', or if they instead feature in the grand visions of those with the insight, ability, charisma and determination to successfully shape the future.

TECHNOLOGICAL DETERMINISM?

As I write the final pages of this book, I am in the closing weeks of a 25 year academic career. During my time in academia I have pondered many theories, the most interesting of which has been 'technological determinism'. This is the view that a civilization's social, cultural and economic structures are determined by its technologies. Technological determinists subsequently believe that the invention of significant new technologies will drive radical change. A technological determinist would, for example, argue that the technologies of the Industrial Revolution created industrial civilization, and that more recently the invention of the Internet fundamentally changed the world.

Technological determinism would hardly be an interesting theory if everybody believed it to be correct. Academics are, after all, a diverse bunch of individuals, with many in my field taking the view that there is no such thing as technological determinism. Rather, they would contend, technology is at best a tool of revolution, with economic, social and cultural forces being the true drivers of any great period of transformation. How, after all, can a technology possibly drive change when it is not a conscious entity or 'social actor'? According to those who oppose the 'reductionist' view of technological determinism, it really has to be people – via their economic, social or cultural actions – who change the world, with technologies created or embraced as needs dictate.

The debate surrounding technological determinism is one of those rare academic arguments with a real, practical relevance. If, after all, a technology can be deterministic, then it becomes critically important for all of those individuals and businesses who may be impacted by it to embrace its revolution as rapidly as possible. On the other hand, if no technology can ever be deterministic, then it is safe for people and companies to ignore new technologies until economic, cultural or social forces indicate that they may need to be adopted.

Personally I take the view that, most of the time, the vast majority of technologies are not deterministic. To the dismay of manufacturers, the creation of most new technologies – such as 3D TVs or smartwatches – therefore does not change the world. There are, however, rare occasions when a technology offers such radical possibilities for new ways of living that it becomes deterministic for a relatively short period. The Internet, for example, I believe was a highly deterministic technology from the early nineties to the late noughties, as during this period its mass adoption engendered extraordinary changes in both our social structures and the business world. Nobody planned for this to happen. Indeed, even some successful computing companies – including Microsoft – came late to what in 1995 Bill Gates termed 'the Internet tidal wave'.

The Last Big Thing of the Internet Revolution was clearly larger than any one company, industry or country. To me at least, this lends weight to the idea that the technology itself was attractive and radical enough to actually create change. This said, given that the Internet emerged out of work at the Defense Advanced Research Projects Agency (DARPA) in the United States in the 1960s, it should be clear that the Internet was not immediately a deterministic technology. Nor, as I argued in the *Prologue*, is the Internet a deterministic technology today.

Looking to the future, it is both fascinating and significant to ponder just which new technologies are destined to become deterministic and when. Such a forecast could clearly help both individuals and organizations to plan ahead. Anticipating the next technological tidal wave has also been a major goal of this book. So, as we approach the final pages, where exactly do things sit?

Well, my own prediction is that artificial intelligence will be the next technology to enter a deterministic phase, with this likely to occur in the 2020s. This, I would suggest, is going to happen because no industry will remain untouched by the wave of mental automation that AI will permit. Nor will many companies remain unaffected by the opportunity to use AI to mine Big Data and to remove language barriers.

In the form of cloud-based virtual assistants like Cortana, Siri, Google Now or Facebook's 'M', AI is also very likely to become the next customer interface. In the same way that most companies had to embrace the world-wide web, so they will subsequently have to embrace AI. In addition to all of this, AI has the power to catalyze the development of every other technology featured in this book, including all forms of digital fabrication and genetic manipulation.

As AI enters its deterministic phase, in parallel I expect humanoid robots to become a considerable driver of industrial and social change. AI and humanoid robots may even remain deterministic for a longer period than any other technology in history. This is because AIs and robots have the potential to become 'social actors' who are capable of taking conscious decisions to change the world.

In the longer term, the other technology that I expect to become technologically deterministic is microfabrication. This I expect to happen in the 2030s, by which time 3D printing, synthetic biology and nanotechnology will have significantly converged. This means that, in around 20 years, microfabricator hardware will be able to locally digitally manufacture a wide spectrum of highly complex inorganic and organic products in pretty much any location.

To be clear, I do not expect 3D printing in its current form to become technologically deterministic. Nor do I expect synthetic biology or nanotechnology 2.0 by themselves to engender radical social, cultural or economic change. Yet, as and when these three fields of digital fabrication fundamentally converge, so the very mechanisms of capitalism will be challenged as the means of production are placed into the hands of the majority.

Today it is perfectly possible for a private citizen to download an object from the Internet and to print it out on a $300 3D printer. This amazing point noted, today it is not possible to personally 3D print the vast majority of the products that populate our lives. Yet in two decades time, I think it is reasonable to predict that the things we will be able to download and fabricate will be common everyday stuff.

EMBRACING CHANGE

AI and robots are now fairly close to their likely period of technologically deterministic takeoff. This means that it is time for those with vision to start embracing their future potential. It is also already soon enough for individuals or businesses seeking a head start in microfabrication to begin experimenting with 3D printing. Almost certainly, 3D printing will be the first local digital manufacturing technology to enter mainstream application, with niche sectors of the aerospace, medical and other industries already starting to be transformed.

The early adopters of personal computers had desktop hardware in their workplaces and homes a good ten years before the rest of the population. It is therefore time for tomorrow's leaders in microfabrication to start building their expertise. There seems little doubt that a digital transition from the age of the microprocessor to that of the microfabricator is going to occur. In turn this means that a great many people will need to become familiar with the creation, manipulation and exchange of digital 3D content. Such skills can also start to be honed right now using today's 3D printing hardware, and indeed today's free 3D software applications.

By now you may be wondering about the significance of the content covered in the last five chapters of this book. I have, after all, just predicted that only those technologies included in the first five chapters will become deterministic! Well, it remains the case that technologies may help to shape the future even if they never enter their own unique deterministic phase. There was, for example, never a period in history when electric light bulbs, antibiotics or the internal combustion engine proved themselves to be technologically deterministic. Nevertheless, all of these technologies had a radical impact on our civilization. It is just that they took off and evolved in their application over a very long period of time, and in parallel with many other developments.

Some great periods of transformation stem from the efforts of a few visionary individuals. I also suspect that it will be amazing individuals who will lead and inspire our species to obtain resources from space. No particular rocket, solar power satellite or mining craft is going to be technologically deterministic. Rather, I imagine that a few great leaders will catalyze the human race to industrially reach beyond our first planet.

Sometime this century, the collective survival instinct to secure a sustainable supply of future resources is also likely to drive developments in space-based solar power and extraterrestrial mining. In turn, this will I suspect trigger the genetic and cybernetic upgrade of some human beings in order to facilitate our part-migration into space.

As yet another force for radical change, over the coming decades I expect that the social, cultural and potentially political movement of transhumanism will grow significantly. Fuelled by both a selfish lust for an improved life, coupled with a desire to improve the lot of their offspring, people will increasingly demand medical upgrade technologies. Postgenomic medicine and advanced cybernetics are I imagine unlikely to ever become technologically deterministic. But this is simply because the forces of social and cultural determinism will pull these technologies into the mainstream long before a period of technological determinism has the opportunity to arise.

SINGULAR GRAND VISION

Any civilization is founded on two pillars. The first of these is its technologies, or in other words the civilization's mechanisms for fabrication and otherwise physically realizing its ambitions. The second pillar of any civilization is then its collective narrative, or in other words those visions, stories and systems of belief that encapsulate its purpose.

Across history, the most successful, momentous undertakings have resulted from a combination of great technologies and great narratives. For example, the Ancient Egyptians leveraged their construction capabilities to build pyramids that fulfilled the narrative of immortalizing their pharaohs. Meanwhile, in the 1960s, NASA created space travel technologies that allowed a nation to realize its narrative vision to land on the Moon. Great technologies demand great ideas to which many citizens commit. And when the two coincide, the world is quite likely to change.

Throughout this book I have made reference to the Singularity. This is a future moment in history that we will potentially reach when almost anything becomes technologically possible. Or as I argued in the *Prologue*, we will have reached the Singularity when the divide between technology and magic begins to blur.

Our arrival at the Singularity will require future advancements in fields including AI and microfabrication. Yet I suspect that we will be unable to reach the Singularity via technological advancement alone. Indeed, were this to be possible, the implication would be that our journey toward the Singularity is solely dependent on a process of technological determinism, and I do not believe this to be the case.

Arriving at a point in future history where the impossible becomes possible will allow us to rise to every grand challenge that confronts the human race. In fact, if we fail to reach the Singularity, then our alternative future is almost inevitably one of relentless decline due to ever-diminishing natural resources. We really do need to develop microfabrication and AI, to reach out for resources from space, and in the process to become transhuman, if our civilization is to survive and thrive. Yet I suspect that all of these things will only ever be achieved if we collectively commit to a bold, new narrative that champions the attainment of such goals. In other words, we all need to develop a belief in the importance of arriving at the Singularity if we are ever to actually get there. And clearly today such a collective belief does not exist.

It is, in fact, difficult to argue that any strong, collective, future-shaping belief exists on this planet. Granted, some religions have a very powerful hold on the minds of their followers. But sadly this power is rarely wielded with an effective future shaping intent. Rather, like most governments, religions fight hard to maintain the status quo. It is indeed for this reason that transhumanism and many traditional religions are going to clash in the decades ahead.

Away from legacy cultural and social structures, a realization is growing that things must change. In particular, I believe that at least some people now recognize that the technologies and related undertakings showcased in this book represent our last, great hope for sustainable, long-term survival. Going back to living in caves is not an option for seven billion people. As on so many other occasions across history, we therefore need to rise to the challenge of tough times ahead by expanding our horizons rather than beating a retreat.

Without a powerful, collective intent to journey toward the Singularity, there is a risk that many of the Next Big Things that we have looked at in this book will rise and fall as passing fads. It is therefore important for as many people as possible to develop an aspirational attitude toward new technology application in the decades ahead.

REFERENCES

Prologue: **Embracing the Singularity**

Christopher Barnatt *Cyber Business: Mindsets for a Wired Age* (Chichester: John Wiley & Sons, 1995).

Arthur C. Clarke *Profiles of The Future* (1961). Last re-published by The Orion Publishing Group Ltd, 2000.

Susumu Sasaki 'How Japan Plans to Build an Orbital Solar Farm', *IEEE Spectrum* (24 April 2014). Available from: http://spectrum.ieee.org/green-tech/solar/how-japan-plans-to-build-an-orbital-solar-farm

Chapter 1: 3D Printing

3Ders.org *3D Printed Titanium Components Now Onboard The Airbus A350 XWB* (21 October 2014). Available from: http://www.3ders.org/articles/20141021-additively-manufactured-titanium-components-now-onboard-the-airbus-a350-xwb.html

3Ders.org *Chinese Government Unveils 'National Plan' For Development Of 3D Printing Industry* (2 March 2015). Available from: http://www.3ders.org/articles/20150302-chinese-government-unveils-national-plan-for-development-of-3d-printing-industry.html

3Dders.org *Deloitte Predicts 220,000 3D Printers Sold In 2015, But Says Money Is In Professional Grade 3D Printing* (2 October 2015). Available from: http://www.3ders.org//articles/20151002-deloitte-predicts-3d-printers-money-is-in-professional-grade-3d-printing.html

3Ders.org *Next Eurostar E3000 Satellite To Feature Airbus 3D Printed Aluminum Parts* (21 March 2015). Available from http://www.3ders.org/articles/20150321-next-eurostar-e3000-satellite-to-feature-airbus-3d-printed-aluminum-parts.html

3Ders.org *WinSun China Builds World's First 3D Printed Villa And Tallest 3D Printed Apartment Building* (18 January 2015). Available from: http://www.3ders.org/articles/20150118-winsun-builds-world-first-3d-printed-villa-and-tallest-3d-printed-building-in-china.html

Gary Anderson '3D Printed Circuit Boards with Graphene Filament from Graphene 3D Lab', *3DPrintingStocks.com* (17 September 2015). Available from: https://3dprintingstocks.com/3d-printed/

Christopher Barnatt *3D Printing: Second Edition* (ExplainingTheFuture.com, 2014).

ExOne *Case Studies*. Available from: http://www.exone.com/Resources/Case-Studies

Fab Lab FAQ. Available from: http://fab.cba.mit.edu/about/faq/

Yu Kyung Jung & Sang Yup Lee 'Efficient Production Of Polylactic Acid And Its Copolymers by Metabolically Engineered Escherichia Coli', *Journal of Biotechnology* (Volume 151, 2011, pp.94–101)

Brian Krassenstein 'ORNL 3D Prints Working Shelby Cobra Replica', *3DPrint.com* (10 January 2015). Available from: http://3dprint.com/36433/3d-printed-shelby-cobra/

REFERENCES

Made in Space *Off-World Manufacturing is Here* (Press Release, 19 September 2014). Available from: http://www.madeinspace.us/offworld-manufacturing-is-here

Michael K. Jungling, Patrick A. Wood & Yukihiro Koike *MedTech 3D Printing – A Solution for Innovation?* (Morgan Stanley, 5 September 2013).

Organovo *L'Oreal USA Announces Research Partnership with Organovo to Develop 3-D Bioprinted Skin Tissue* (Press Release, 5 May 205). Available from: http://ir.organovo.com/news/press-releases/press-releases-details/2015/LOreal-USA-Announces-Research-Partnership-with-Organovo-to-Develop-3-D-Bioprinted-Skin-Tissue/

Organovo *Organovo Announces Commercial Release of the exVive3D™ Human Liver Tissue* (Press Release, 18 November 2014). Available from: http://ir.organovo.com/news/press-releases/press-releases-details/2014/Organovo-Announces-Commercial-Release-of-the-exVive3D-Human-Liver-Tissue/

Smithers Pira *Global Market For 3d Printing Expected To Reach Over $49 Billion By 2025* (June 2015). Available via: http://www.smitherspira.com/news/2015/june/3d-print-market-expected-to-reach-$49b-by-2025

SpaceX *SpaceX Completes Qualification Testing of SuperDraco Thruster*, (Press Release, 27 May 2014). Available from: http://www.spacex.com/press/2014/05/27/spacex-completes-qualification-testing-superdraco-thruster

SpaceX *Space X Launches 3D Printed Part to Space, Creates Printed Engine Chamber* (Press Release, 31 July 2014). Available from: http://www.spacex.com/news/2014/07/31/spacex-launches-3d-printed-part-space-creates-printed-engine-chamber-crewed

Stratasys (2015) *Volvo Trucks Slashes Manufacturing Tool Production Time By More Than 94% While Increasing Plant Efficiency With Stratasys 3D Printing* (Press Release, 18 March 2015). Available from: http://www.bespoke.co.uk/index.php?option=com_content&view=article&id=4707

Chapter 2: Synthetic Biology

Algix website: http://algix.com

Amyris Biotechnologies website: https://amyris.com/

Rachel Armstrong *Living Architecture: How Synthetic Biology Can Remake Our Cities and Reshape Our Lives* (TED Books, 2012).

Christopher Barnatt *25 Things You Need to Know About the Future* (London: Constable, 2012).

Dominic Basulto 'The Big Trends in Synthetic Biology You Need to Know', *The Washington Post* (8 October 2015). Available from: https://www.washingtonpost.com/news/innovations/wp/2015/10/08/the-big-trends-in-synthetic-biology-you-need-to-know/

BCC Research *Global Market for Synthetic Biology to Reach Nearly $11.9 Billion by 2018* (Press Release, July 14 2014). Available from: http://www.bccresearch.com/pressroom/bio/global-market-synthetic-biology-reach-nearly-$11.9-billion-2018

Biotech Industry Association *Current Uses of Synthetic Biology for Renewable Chemicals, Pharmaceuticals, and Biofuels* (May 2014). Available via: https://www.bio.org/articles/current-uses-synthetic-biology-renewable-chemicals-pharmaceuticals-and-biofuels

Jerome Bonnet, Peter Yin P, Monica E. Ortiz, Pakpoom Subsoontorn & Dew Endy 'Amplifying Genetic Logic Gates', *Science* (3 May 2013, Vol. 340 no. 6132 pp. 599-603).

D. Ewen Cameron, Caleb J. Bashor & James J. Collins 'A Brief History of Synthetic Biology', Nature Reviews Microbiology (Volume 14, May 2014). Available from: http://collinslab.mit.edu/files/nrm_cameron.pdf

Dickson Despommier *The Vertical Farm: Feeding the World in the 21st Century* (New York, NY: Thomas Dune, 2010).

ERASynBio *Next Steps For A European Synthetic Biology: A Strategic Vision From ERASynBio* (April 2014). Available from: http://www.bbsrc.ac.uk/documents/1404-era-synbio-strategic-vision-pdf/

Field Test Film Corps *Synthetic Bio Documentary: Transgenic Spidergoats Brief* (2010). Video viewable at: https://vimeo.com/17556768

GenoCAD website: http://www.genocad.org/

Jessica Griggs 'Creating Buildings That Repair Themselves', *New Scientist* (22 February 2012). Available from: http://www.newscientist.com/blogs/culturelab/2012/02/creating-buildings-that-repair-themselves.html

iGEM Registry of Standard Biological Parts. Available at: http://parts.igem.org/

Intrexon *Synthetic Biology*. Available from: http://www.dna.com/Company/Synthetic-Biology-History

J. Craig Venter Institute *First Self-Replicating, Synthetic Bacterial Cell Constructed by J. Craig Venter Institute Researchers*. (Press release, May 20 2010). Available from: http://www.jcvi.org/cms/fileadmin/site/research/projects/first-self-replicating-bact-cell/press-release-final.pdf

Life Technologies *Synthetic Biology Applications*. Available from: http://www.lifetechnologies.com/uk/en/home/life-science/synthetic-biology/synthetic-biology-applications.html

Randall Mayes 'Where Will the Century of Biology Lead Us?' *The Futurist* (May-June 2014).

Jacques Monod & Francois Jacob 'Teleonomic Mechanisms In Cellular Metabolism, Growth, And Differentiation' *Cold Spring Harb. Symp. Quant. Biol.* (Volume 26, pp.389–401, 1961).

Solazyme About the Company: http://solazyme.com/company/

Spiber website: http://www.spiber.se

Synthetic Biology Open Language website. Available at: http://sbolstandard.org/

Synthetic Genomics Inc. *Digitizing Life, Developing Transformative Products, Solving Global Challenges* (2015) Available from: http://www.syntheticgenomics.com/

SyntheticBiology.org website: http://syntheticbiology.org/

David Thorpe 'How China Leads the World in Indoor Farming', *Sustainable Cities Collective* (15 September 2014). Available from: http://sustainablecitiescollective.com/david-thorpe/409606/chinas-indoor-farming-research-feed-cities-leads-world

WorldWatch Institute *Genetically Modified Crops Only a Fraction of Primary Global Crop Production* (VST 123). Available from: http://www.worldwatch.org/node/5950

Chapter 3: Nanotechnology 2.0

David L. Chandler 'Enhancing the Power of Batteries', *MIT News* (20 June 2010). Available from: http://newsoffice.mit.edu/2010/batteries-nanotubes-0621

Anne Clunan & Kirsten Rodine-Hardy 'Nanotechnology in a Globalized World: Strategic Assessments of an Emerging Technology' (PASCC Report 2014 006). Available from: http://www.nps.edu/Academics/Centers/CCC/PASCC/Publications/2014/2014%20006%20Nanotechnology%20Strategic%20Assessments.pdf

Larry Dignan 'IBM Research builds functional 7nm processor', *ZDNet* (9 July 2015). Available from: http://www.zdnet.com/article/ibm-research-builds-functional-7nm-processor/

Eric K. Drexler *Engines of Creation: The Coming Era of Nanotechnology* (New York: Anchor Books, 1986). Full text available from: http://e-drexler.com/d/06/00/EOC/EOC_Table_of_Contents.html

Eric K. Drexler *Radical Abundance: How a Revolution in Nanotechnology Will Change Civilization* (New York: Public Affairs, 2013).

Thomas Frey *Communicating with the Future: How Re-engineering Intentions Will Alter the Master Code of Our Future* (Colorado: DaVinci Institute Press, 2011).

FutureTimeline.net *Self-assembling Material Could Lead To Artificial Veins* (3 October 2015). Available from: http://www.futuretimeline.net/blog/2015/10/3.htm#.Vh4xgexVhHz

REFERENCES

Tim Harper '2015: The Year of the Trillion Dollar Nanotechnology Market?', *AZoNano.com* (January 2 2015). Available from: http://www.azonano.com/article.aspx?ArticleID=3946#ref1

William Herkewitz 'This Chemistry 3D Printer Can Synthesize Molecules From Scratch', *Popular Mechanics* (12 March 2015). Available from: http://www.popularmechanics.com/science/health/a14528/the-chemistry-3d-printer-can-craft-rare-medicinal-molecules-from-scratch/

Interagency Working Group on Nanoscience, Engineering and Technology *National Nanotechnology Initiative: Leading to the Next Industrial Revolution.* (Committee on Technology National Science and Technology Council, February 2000). Available from: http://www.whitehouse.gov/files/documents/ostp/NSTC%20Reports/NNI2000.pdf

Ion Publishing *Nanotechnology: Applications and Markets* (2013). Available for purchase via: http://www.nanomagazine.co.uk/

Dexter Johnson 'Graphene Finds Its Path in Supercapacitor Commercialization', *IEEE Spectrum* (3 April 2015). Available from: http://spectrum.ieee.org/nanoclast/semiconductors/materials/graphene-finds-its-path-in-supercapacitor-commercialization

Ralph Merkle *An Introduction to Molecular Nanotechnology* (Singularity University, 2009). Available as a video at: https://www.youtube.com/watch?v=cdKyf8fsH6w

Ron Mertens 'XG Sciences Launches New, High-capacity Graphene-based Anode Materials for Li-Ion Batteries', *Graphene-Info* (11 April 2013). Available from: http://www.graphene-info.com/xg-sciences-new-high-capacity-graphene-based-anode-materials-li-ion-batteries

NSTC/CoT/NSET *NNI Supplement to the President's 2015 Budget* (NNI Publications and Reports, 25 March 2014). Available from: http://www.nano.gov/node/1128

Steven J. Oldenburg 'Silver Nanoparticles: Properties and Applications', *Sigma-Aldrich* (2015). Available from: http://www.sigmaaldrich.com/materials-science/nanomaterials/silver-nanoparticles.html

QMUL *Self-assembling Material That Grows And Changes Shape Could Lead To Artificial Arteries* (Press Release, 28 September 2015). Available from: http://www.qmul.ac.uk/media/news/items/se/164051.html#

Darren Quick 'Graphene-based Solar Cell Hits Record 15.6 Percent Efficiency', *Gizmag* (14 January 2014). Available from: http://www.gizmag.com/graphene-solar-cell-record-efficiency/30466/

Alexander Sinitskii & James M. Tour 'Graphene Electronics, Unzipped', *IEEE Spectrum* (29 October 2010). Available from: http://spectrum.ieee.org/semiconductors/materials/graphene-electronics-unzipped

Ryan Smith 'Intel's 14nm Technology in Detail' *AnandTech* (11 August 2014). Available from: http://www.anandtech.com/show/8367/intels-14nm-technology-in-detail

John Sullivan *Nanotechnology Leads To Better, Cheaper Leds For Phones And Lighting* (Princeton University, 24 September 2014). Available from: http://www.princeton.edu/main/news/archive/S41/14/79S63/

The Project for Emerging Nanotechnologies *Consumer Products Inventory* (2015). Available at: http://www.nanotechproject.org/cpi/

University of Illinois *Molecule-making Machine Simplifies Complex Chemistry* (Press Release, 12 March 2014). Available from: http://news.illinois.edu/news/15/0312molecule_machine_MartyBurke.html

Chapter 4: Artificial Intelligence

Association for the Advancement of Artificial Intelligence website: http://www.aaai.org

Audi Piloted Driving web pages at: http://www.audi.com/content/com/brand/en/vorsprung_durch_technik/content/2014/10/piloted-driving.html

CNN Money *Could This Computer Save Your Life?* (12 March 2015): Available from: http://money.cnn.com/2015/03/12/technology/enlitic-technology/

Rory Cellan-Jones 'Stephen Hawking Warns Artificial Intelligence Could End Mankind', *BBC News* (2 December 2015). Available from: http://www.bbc.co.uk/news/technology-30290540

Cisco *Cisco Global Cloud Index: Forecast and Methodology, 2013–2018* (White Paper, 2014). Available from: http://www.cisco.com/c/en/us/solutions/collateral/service-provider/global-cloud-index-gci/Cloud_Index_White_Paper.html

Cognitec website: http://www.cognitec.com/

CNN Money *Facebook's New App Powered By Artificial Intelligence* (15 June 2015). Available from: http://money.cnn.com/2015/06/15/technology/facebook-moments-ai/

D-Wave Systems website: http://www.dwavesys.com/

Daimler *Freightliner Inspiration Truck World Premiere on Hoover Dam* (Press Release, 5 May 2015). Available from: http://www.daimler.com/dccom/0-5-1809607-1-1809608-1-0-0-0-0-0-0-0-0-0-0-0-0-0.html

Alex Davies 'The UK Just Made Itself a Fantastic Place to Test Self-Driving Cars', *Wired* (12 February 2015). Available from: http://www.wired.com/2015/02/uk-just-made-fantastic-place-test-self-driving-cars/

Enlitic website: http://www.enlitic.com/

Facebook AI Research website: https://research.facebook.com/ai

Google Self Driving Car website: http://www.google.com/selfdrivingcar/

Google Translate Blog: http://googletranslate.blogspot.co.uk/

Andy Greenberg 'After Jeep Hack, Chrysler Recalls 1.4M Vehicles for Bug Fix', *Wired* (25 July 2015). Available from: http://www.wired.com/2015/07/jeep-hack-chrysler-recalls-1-4m-vehicles-bug-fix/

Stacey Higginbotham 'How Facebook Is Teaching Computers To See', *Fortune* (15 June 2015). Available from: http://fortune.com/2015/06/15/facebook-ai-moments/

IBM *IBM Forms New Watson Group to Meet Growing Demand for Cognitive Innovations* (News Release, 9 January 2014). Available from: https://www-03.ibm.com/press/us/en/pressrelease/42867.wss

IBM Watson website: http://www.ibm.com/smarterplanet/us/en/ibmwatson/

Ray Kurzweil 'Rise Of The Robots: How Long Do We Have Until They Take Our Jobs?' *The Guardian* (4 February 2015). Available from: http://www.theguardian.com/technology/2015/feb/04/rise-robots-artificial-intelligence-computing-jobs

James Manyika, Michael Chui, Brad Brown, Jacques Bughin, Richard Dobbs, Charles Roxburgh & Angela Hung Byers *Big Data: The Next Frontier for Innovation, Competition, and Productivity*, (McKinsey Global Institute, May 2011).

John Markoff 'Computer Wins on "Jeopardy!": Trivial, It's Not', *New York Times* (16 February 2011). Available from: http://www.nytimes.com/2011/02/17/science/17jeopardy-watson.html

Microsoft Research *Quantum Computing: Creating A New Generation Of Computing Devices*. Available from: http://research.microsoft.com/en-us/research-areas/quantum-computing.aspx

Ron Miller 'IBM's New Watson Analytics Wants To Bring Big Data To The Masses', *Tech Crunch* (16 September 2014). Available from: http://techcrunch.com/2014/09/16/ibms-new-watson-analytics-wants-to-bring-data-to-the-masses/

Brad Power 'Artificial Intelligence Is Almost Ready for Business', *Harvard Business Review* (19 March 2015). Available from: https://hbr.org/2015/03/artificial-intelligence-is-almost-ready-for-business

Quick Response CCTV Ltd *CCTV Facts & Numbers*. Available from: http://qrcctv.co.uk/cctv/cctv-facts

Tom Simonite 'IBM Shows Off a Quantum Computing Chip', *MIT Technology Review* (29 April 2015). Available from: http://www.technologyreview.com/news/537041/ibm-shows-off-a-quantum-computing-chip/

Nick Statt 'Bill Gates Is Worried About Artificial Intelligence Too' *CNET* (28 January 2015). Available from: http://www.cnet.com/news/bill-gates-is-worried-about-artificial-intelligence-too/

The Economist 'Rise of the Machines' (9 May 2015). Available from: http://www.economist.com/news/briefing/21650526-artificial-intelligence-scares-peopleexcessively-so-rise-machines

Tesla Motors Team *Your Autopilot Has Arrived*. (Blog post, 14 October 2015). Available from: http://www.teslamotors.com/blog/your-autopilot-has-arrived

Alan M. Turing 'Computing Machinery and Intelligence'. *Mind*, (Volume LIX, Number 236, October 1950). Now available from multiple sources, including: http://loebner.net/Prizef/TuringArticle.html

University of Reading *Turing Test Success Marks Milestone In Computing History* (Press Release, 8 June 2014). Available from: http://www.reading.ac.uk/news-and-events/releases/PR583836.aspx

James Vlahos 'Surveillance Society: New High-Tech Cameras Are Watching You', *Popular Mechanics* (30 September 2009). Available from: http://www.popularmechanics.com/military/a2398/4236865/

Jane Wakefield 'Driverless Car Review Launched By Uk Government', *BBC News* (15 February 2015). Available from: http://www.bbc.co.uk/news/technology-31364441

Gillian Yeomans *Autonomous Vehicles: Handing Over Control: Opportunities and Risks For Insurance* (Lloyds, 2014). Available from: https://www.lloyds.com/~/media/lloyds/reports/emerging%20risk%20reports/autonomous%20vehicles%20final.pdf

Linda Yueh 'Can Huawei become China's first global brand?', *BBC News* (17 October 2014). Available from: http://www.bbc.co.uk/news/business-29628044

James Condliffe 'China Has the World's First Face Recogniton ATM', Gizmodo (6 January 2015). Available from: http://gizmodo.com/china-has-the-worlds-first-facial-recognition-atm-1708132476

Chapter 5: Humanoid Robots

Evan Ackerman 'NASA JSC Unveils Valkyrie DRC Robot', *IEEE Spectrum* (10 December 2013). Available from: http://spectrum.ieee.org/automaton/robotics/military-robots/nasa-jsc-unveils-valkyrie-drc-robot

Aldebaran website: https://www.aldebaran.com/en

Boston Consulting Group R*ise of the Machines: BCG Projects $67 Billion Market for Robots by 2025* (Press release, 3 September 2014). Available from: http://www.bcg.com/news/press/3sep2014-rise-machines-2025.aspx

Boston Consulting Group *Takeoff in Robotics Will Power the Next Productivity Surge in Manufacturing* (Press Release, 10 February 2015). Available from: http://www.bcg.com/news/press/10feb2015-robotics-power-productivity-surge-manufacturing.aspx

Boston Dynamics website: http://www.bostondynamics.com/

Marshall Brain *The Second Intelligent Species: How Humans Will Become as Irrelevant as Cockroaches*. Available via: http://marshallbrain.com/second-intelligent-species.htm

Marshall Brain *Robotic Nation*. Available at: http://marshallbrain.com/robotic-nation.htm

DARPA Robotics Challenge website: http://www.theroboticschallenge.org/

Megan Rose Dickey 'A Robot Army Is Destroying The American Workforce', *Business Insider* (15 January 2013). Available from: http://www.businessinsider.com/a-robot-army-is-destroying-the-american-workforce-2013-1?IR=T

The Economist 'The Dawn of Artificial Intelligence' (9 May 2015). Available from: http://www.economist.com/news/leaders/21650543-powerful-computers-will-reshape-humanitys-future-how-ensure-promise-outweighs?fsrc=scn/tw/te/pe/ed/AIleader

Martin Ford *Rise of the Robots: Technology and the Threat of a Jobless Future* (New York: Basic Books, 2015).

Erico Guizzo 'How Aldebaran Robotics Built Its Friendly Humanoid Robot, Pepper', *IEEE Spectrum* (26 December 2014). Available from: http://spectrum.ieee.org/robotics/home-robots/how-aldebaran-robotics-built-its-friendly-humanoid-robot-pepper

Erico Guizzo & Evan Ackerman 'How South Korea's DRC-HUBO Robot Won the DARPA Robotics Challenge', *IEEE Spectrum* (9 June 2015). Available from: http://spectrum.ieee.org/automaton/robotics/humanoids/how-kaist-drc-hubo-won-darpa-robotics-challenge

Honda ASIMO webite at: http://asimo.honda.com/

Honda *Honda to Begin Lease Sales of Honda Walking Assist Device in Japan* (Press Release, 21 July 2015). Available from: http://hondanews.com/honda-corporate/channels/robotics-asimo/releases/honda-to-begin-lease-sales-of-honda-walking-assist-device-in-japan

IEEE Spectrum humanoid robot website: http://spectrum.ieee.org/robotics/humanoids

InMoov website: http://www.inmoov.fr/

International Federation of Robotics *World Robotics 2014 Industrial Robots*. Data and Executive Summary available via: http://www.ifr.org/industrial-robots/statistics/

Xeni Jardin 'Elon Musk and Stephen Hawking Call For Ban On "autonomous Weapons"', *BoingBoing* (28 January 2015). Available from: http://boingboing.net/2015/07/28/elon-musk-and-stephen-hawking.html

Masahiro Mori 'The Uncanny Valley', translation by Karl F. MacDorman and Norri Kageki, *IEEE Spectrum* (12 June 2012). Available from: http://spectrum.ieee.org/automaton/robotics/humanoids/the-uncanny-valley

Michael Molitch-Hou 'Hospitalized Children to Receive Virtual Zoo Tours from the 3D Printed InMoov Robot', *3DPrintingIndustry.com* (15 Jan 2015). Available from: http://3dprintingindustry.com/2015/01/15/hospitalized-children-receive-virtual-zoo-tours-3d-printed-inmoov-robot/

NASA *NASA Looks to University Robotics Groups to Advance Latest Humanoid Robot* (Press Release, 11 June 2015). Available from: http://www.nasa.gov/feature/nasa-looks-to-university-robotics-groups-to-advance-latest-humanoid-robot

Max Nisen 'Robot Economy Could Cause Up To 75 Percent Unemployment', *Business Insider* (28 January 2013). Available from: http://www.businessinsider.com/50-percent-unemployment-robot-economy-2013-1?IR=T

Poppy Project website: https://www.poppy-project.org/

Robots for Good website: http://www.robotsforgood.com/

Project Romeo website: http://projetromeo.com/en

Steve Rose 'Ex Machina And Sci-fi's Obsession With Sexy Female Robots', *The Guardian* (15 January 2015). Available from: http://www.theguardian.com/film/2015/jan/15/ex-machina-sexy-female-robots-scifi-film-obsession

Alison Sander & Meldon Wolfgang 'The Rise of Robotics', *BCG Perspectives* (27 August 2014). Available from: https://www.bcgperspectives.com/content/articles/business_unit_strategy_innovation_rise_of_robotics/

Angad Singh '"Emotional" Robot Sells Out in a Minute', *CNN* (23 June 2015). Available from: http://edition.cnn.com/2015/06/22/tech/pepper-robot-sold-out/

SoftBank Robotics Corp *SoftBank to Take Orders for Pepper Enterprise Model - 'Pepper for Biz' - Starting October 1* (Press Release, 30 July 2015). Available from: http://www.softbank.jp/en/corp/group/sbr/news/press/2015/20150730_01/

Singularity Weblog *Martin Ford on Singularity 1on1: Technological Unemployment is an Issue We Need To Discuss* (13 June 2015). Available from: https://www.singularityweblog.com/martin-ford-on-technological-unemployment/

John Tamny 'Why Robots Will Be The Biggest Job Creators in History', *Forbes* (1 March 2015). Available from: http://www.forbes.com/sites/johntamny/2015/03/01/why-robots-will-be-the-biggest-job-creators-in-history/

George Webster 'Print Your Own Life-size Robot For Under $1,000' *CNN* (25 January

2015). Available from: http://edition.cnn.com/2013/01/25/tech/innovation/inmoov-robot-3d-printing/

Chapter 6: Space-Based Solar Power

Evan Ackerman 'Japan Demos Wireless Power Transmission for Space-Based Solar Farms', *IEEE Spectrum* (16 March 2015). Available from: http://spectrum.ieee.org/energywise/green-tech/solar/japan-demoes-wireless-power-transmission-for-spacebased-solar-farms

Airbus Defence & Space *Space Based Solar Power*. Available from: http://www.space-airbus-ds.com/en/programmes/space-based-solar-power-innovating-for-clean-energy-k0f.html

Jonathan Amos 'EADS Astrium Develops Space Power Concept', *BBC News* (19 January 2010). Available from: http://news.bbc.co.uk/1/hi/sci/tech/8467472.stm

Boeing Solar Power Satellite web page at: http://www.boeing.com/history/products/solar-power-satellite.page

Marc Boucher 'Is There a Future for Space-Based Solar Power?', *Space Ref* (11 June 2012). Available from: http://spaceref.com/space-quarterly-magazine/is-there-a-future-for-space-based-solar-power.html

William C. Brown 'The Microwave Powered Helicopter', *The Journal of Microwave Power* (Volume 1 Number 1, 1964). Available from: http://www.jmpee.org/JMPEE_PDFs/01-1_bl/JMPEE-Vol1-Pg1-Brown.pdf

Eric K. Drexler *Radical Abundance: How a Revolution in Nanotechnology Will Change Civilization* (New York: Public Affairs, 2013).

English.News.cn 'Chinese Scientists Mull Power Station in Space' (30 March 2015). Available from: http://news.xinhuanet.com/english/2015-03/30/c_134109115.htm

Loren Grush 'How Spacex And Boeing Plan To Keep Nasa Astronauts Safe', *The Verge* (13 August 2015). Available from: http://www.theverge.com/2015/8/13/9142471/spacex-boeing-plan-to-keep-nasa-astronauts-safe

Loretta Hidalgo 'Researchers Beam "space" Solar Power In Hawaii', *Wired* (12 September 2008). Available from: http://www.wired.com/2008/09/visionary-beams/

Japan Space Systems *Space Solar Power System*. Available from: http://www.jspacesystems.or.jp/en_project_ssps/

JAXA *Practical Application of Space-Based Solar Power Generation -- interview with Yasuyuki Fukumuro* (8 June 2010). Available from: http://global.jaxa.jp/article/interview/vol53/index_e.html

Lasermotive *Laser Power Beaming Fact Sheet*. Available from: http://lasermotive.com/wp-content/uploads/2012/03/Laser-Power-Beaming-Fact-Sheet.pdf

John C. Mankins (ed) *The First International Assessment Of Space Solar Power: Opportunities, Issues And Potential Pathways Forward* (International Academy of Astronautics, 2001). Available from: http://iaaweb.org/iaa/Studies/sg311_finalreport_solarpower.pdf

John C. Mankins *The Case for Solar Power* (Virginia Edition Publishing, 2014). Available from: http://aeweb.tamu.edu/aero489/ESBI.Spring.15/the%20case%20for%20solarpower.pdf

John C. Mankins *SPS-ALPHA: The First Practical Solar Power Satellite via Arbitrarily Large Phased Array* (NASA, 15 September 2012). Available from: http://www.nasa.gov/pdf/716070main_Mankins_2011_PhI_SPS_Alpha.pdf

Donella H. Meadows, Dennis L. Meadows, Jorgen Randers & William W. Behrens III *The Limits to Growth: A Report for the Club of Rome's Project on the Predicament of Mankind* (New York: Universe Books, 1972).

Donella H. Meadows, Dennis L. Meadows & Jorgen Randers *Beyond The Limits: Global Collapse or a Sustainable Future* (London: Earthscan Publications, 1992).

Donella H. Meadows, Jorgen Randers & Dennis L. Meadows *Limits to Growth: The 30-Year Update* (White River Junction, Vermont: Chelsea Green Publishing, 2002).

Space Energy website at: http://www.spaceenergy.com

Susumu Sasaki 'How Japan Plans to Build an Orbital Solar Farm', *IEEE Spectrum* (24 April 2014). Available from: http://spectrum.ieee.org/green-tech/solar/how-japan-plans-to-build-an-orbital-solar-farm

World Commission on Environment and Development Our Common Future (Oxford: Oxford University Press, 1987). Available from: http://www.un-documents.net/our-common-future.pdf

Chapter 7: Asteroid Mining

Asterank website at: http://www.asterank.com/

Brad R. Blair *The Role of Near-Earth Asteroids in Long-Term Platinum Supply* (5 May 2000). Available from: http://www.nss.org/settlement/asteroids/RoleOfNearEarthAsteroidsInLongTermPlatinumSupply.pdf

Deep Space Industries website at: http://deepspaceindustries.com/

JAXA Hayabusa web page at: http://hayabusa.jaxa.jp/e/

Kick Institute for Space Studies *Asteroid Retrieval Feasibility Study* (2 April 2012). Available from: http://www.kiss.caltech.edu/study/asteroid/asteroid_final_report.pdf

John S. Lewis *Asteroid Mining 101: Wealth for the New Space Economy* (Amazon Digital Services, 2015).

Clara Moskowitz 'Giant Asteroid Vesta Surprisingly Covered in Hydrogen', *Space.com* (20 September 2012). Available from: http://www.space.com/17680-giant-asteroid-vesta-hydrogen-dawn-spacecraft.html

NASA Asteroid Redirect Mission web pages at: http://www.nasa.gov/mission_pages/asteroids/initiative/index.html

NASA's Orion website at: http://www.nasa.gov/exploration/systems/orion/index.html

Planetary Resources website at: http://www.planetaryresources.com/

Shane D. Ross *Near-Earth Asteroid Mining* (14 December 2001). Available from: http://2004.isdc.nss.org/settlement/asteroids/NearEarthAsteroidMining(Ross2001).pdf

Mark Sonter 'Asteroid Mining: Key to the Space Economy', (National Space Society, February 2006). Available from: http://www.nss.org/settlement/asteroids/key.html

Andrew Topf '$5-trillion Asteroid To Whiz Past Earth', *Mining.com* (19 July 2015). Available from: http://www.mining.com/5-trillion-asteroid-to-whiz-past-earth/

UNEP *Decoupling Natural Resource Use and Environmental Impacts from Economic Growth* (2011). Available from: http://www.unep.org/resourcepanel/decoupling/files/pdf/Decoupling_Report_English.pdf

Chapter 8: Mining the Moon

Astrobotic website at: https://www.astrobotic.com/

Christopher Barnatt *25 Things You Need to Know About the Future* (London: Constable, 2012).

Dominic Basulto 'Why It Matters That Japan Is Going To The Moon', *Washington Post* (20 April 2015). Available from: http://www.washingtonpost.com/news/innovations/wp/2015/04/30/why-it-matters-that-japan-is-going-to-the-moon/

Fabrizio Bozzato 'The Red Side of the Moon: China's Pursuit of Lunar Helium 3', *eRanli Magazine* (2 July 2014). Available from: http://www.erenlai.com/en/home/item/5892-the-red-side-of-the-moon-china-s-pursuit-of-lunar-helium-3.html

Susan Caminiti 'Billionaire Teams Up With NASA to Mine the Moon' *NBC News* (10 March 2015). Available from: http://www.nbcnews.com/science/space/billionaire-teams-nasa-mine-moon-n321006

T.M. Eubanks *The Acceleration of the Human Exploration of the Solar System with Space Elevators* (Global Space Exploration Conference 2012). GLEX-2012.02.P.2x12186.

Wenzhe Fa and Ya-Qiu Jin, 'Quantitative Estimation of Helium-3 Spatial Distribution

REFERENCES

in the Lunar Regolith Layer', *Icarus* (April 2007, No. 190.

Google Lunar XPRIZE website at: http://lunar.xprize.org/

Richard Hollingham 'Should We Build a Villiage on the Moon?', *BBC Future* (13 July 2015). Available from: http://www.bbc.com/future/story/20150712-should-we-build-a-village-on-the-moon

Lunar Solar Power website at: http://www.lunarsolarpower.org

Euan McKirdy 'Japan's space agency aims for the moon in 2018', *CNN* (14 May 2015). Available from: http://edition.cnn.com/2015/04/23/tech/japan-moon-lander-planned/

Doug Messier 'Shackleton Energy Company Launches Plan for First Lunar Mining Operation', *Parabolic Arc* (November 9, 2011). Available from: http://www.parabolicarc.com/2011/11/09/exclusive-shackleton-energy-company-launches-plan-for-first-lunar-mining-operation/

Charles Miller *et al Economic Assessment and Systems Analysis of an Evolvable Lunar Architecture that Leverages Commercial Space Capabilities and Public-Private-Partnerships* (NexGen Space LLC, 13 July 2015). Available from: http://science.ksc.nasa.gov/shuttle/nexgen/Nexgen_Downloads/NexGen_ELA_Report_FINAL.pdf

Moon Express website at: http://www.moonexpress.com/

NASA's COTS website at: http://www.nasa.gov/offices/c3po/home/cots_project.html

NASA LCROSS website at: http://www.nasa.gov/mission_pages/LCROSS/main/

NASA's Lunar CATALYST website at: http://www.nasa.gov/lunarcatalyst/

NASA's Lunar Flashlight website at: http://sservi.nasa.gov/articles/lunar-flashlight/

NASA's Resource Prospector website at: https://www.nasa.gov/resource-prospector

NASA *NASA Missions Uncover the Moon's Buried Treasures* (Press Release, 21 October 2010). Available from: http://www.nasa.gov/centers/ames/news/releases/2010/10-89AR.html

John O'Ceallaigh 'Mankind Returns to the Moon' *The Daily Telegraph* (6 October 2014). Available from: http://www.telegraph.co.uk/luxury/travel/47469/mankind-returns-to-the-moon.html

Andrew Osborn 'Russia Plans To Put A Mine On The Moon To Help Boost Energy Supply', *The Independent* (2 April 2009). Available from: http://www.independent.co.uk/news/world/europe/russia-plans-to-put-a-mine-on-the-moon-to-help-boost-energy-supply-6110683.html

Austin Ramzy 'China Celebrates Lunar Probe and Announces Return Plans', *The New York Times* (16 December 2013). Available from: http://sinosphere.blogs.nytimes.com/2013/12/16/china-celebrates-lunar-probe-and-announces-return-plans/

John Roach 'Moonbase Announced by NASA' *National Geographic* (4 December 2006). Available from: http://news.nationalgeographic.com/news/2006/12/061204-moon-base.html

Everett Rosenfeld 'Russian Firm Proposes $9.4b Moon Base For Mining', *CNBC* (2 January 2015). Available from: http://www.cnbc.com/2015/01/02/russian-firm-proposes-94b-moon-base-for-mining.html

Shackleton Energy Company website at: http://www.shackletonenergy.com

Sputnik News *Roscosmos Revives Permanent Moon Base Plans* (19 January 2012). Available from: http://www.sputniknews.com/science/20120119/170840782.html

Will Stewart 'Pitin: We Are Coming to the Moon Forever', *Daily Mail* (11 April 2014). Available from: http://www.dailymail.co.uk/news/article-2602291/We-coming-Moon-FOREVER-Russia-sets-plans-conquer-colonise-space-including-permanent-manned-moon-base.html

United Nations Office for Outer Space Affairs *The Outer Space Treaty*. Available from: http://www.unoosa.org/oosa/en/ourwork/spacelaw/treaties/outerspacetreaty.html

Mike Wall 'The Moon's History Is Surprisingly Complex, Chinese Rover Finds', *Space.com* (12 March 2015). Available from: http://www.space.com/28810-moon-history-chinese-lunar-rover.html

Miles Yu 'Russia And China Aim For The Moon And A Joint Lunar Base', *The Washington Times* (20 May 2015). Available from: http://www.washingtontimes.com/news/2015/may/7/inside-china-china-russia-to-build-moon-base/?page=all

Chapter 9: Post-Genomic Medicine

23andMe website at: https://www.23andme.com

Affymetrix website at: http://www.affymetrix.com

Assurex website at: http://assurexhealth.com

Christopher Barnatt *25 Things You Need to Know About the Future* (London: Constable, 2012).

Francie Diep 'Pig Heart Transplants For Humans Are On The Way', *Popular Science* (30 April 2014). Available from: http://www.popsci.com/article/science/pig-heart-transplants-humans-are-way

Fertility Institutes website at: http://www.fertility-docs.com

GeneTherapy.net website at: http://www.genetherapynet.com

Julian Huxley *Religion without Revelation* (The New American Library, 1927. Revised 1956).

Tros de Ilarduya, Sun Y & Düzgüne N. 'Gene Delivery By Lipoplexes And Polyplexes', *Eurpean Journal of Pharmaceutical Science* (Volume 40 No 3, 14 June 2014). Available via: http://www.ncbi.nlm.nih.gov/pubmed/20359532

Illumina website at: http://www.illumina.com

Learn Genetics *Gene Therapy Successes*. Available from: http://learn.genetics.utah.edu/content/genetherapy/gtsuccess/

Alain Li-Wan-Po & Peter Farndon 'Pharmacogenetics: an Introduction', *The Pharmacist*. Available at: http://www.geneticseducation.nhs.uk/downloads/0428PharmacistArticle08.pdf

Elizabeth Lopatto '23andMe Expands To The UK Despite Us Restrictions', *The Verge* (1 December 2014). Available from: http://www.theverge.com/science/2014/12/1/7316089/23andme-expands-to-the-uk-despite-us-restrictions

Max More 'Transhumanism: A Futurist Philosophy', *Extropy* (1990, no.6).

National Human Genome Research Institute website at: http://www.genome.gov/

Nanobiosym website at: http://www.nanobiosym.com

NHS Choices *Gene Therapy Breakthrough For Cystic Fibrosis* (3 July 2015). Available from: http://www.nhs.uk/news/2015/07July/Pages/Gene-therapy-breakthrough-for-cystic-fibrosis.aspx

NHS *Pharmacogenomics in Healthcare*. Available from: http://www.geneticseducation.nhs.uk/pharmacogenomics-in-healthcare

Qualcom Tricorder XPRIZE website at: http://tricorder.xprize.org

Naomi Schaefer Riley '"Designer Babies"' Are An Unregulated Reality, *New York Post* (5 July 2015). Available from: http://nypost.com/2015/07/05/designer-babies-are-an-unregulated-reality/

Paul J. H. Schoemaker and Joyce A. Schoemaker, *Chips, Clones and Living Beyond 100: How Far Will the Biosciences Take Us?* (Upper Saddle River, NJ: FT Press Science, 2010).

Chapter 10

3Ders.org *3D Bone Printing Project In China To Enter Animal Testing Stage* (15 July 2015). Available from: http://www.3ders.org/articles/20150715-3d-bone-printing-project-in-china-to-enter-animal-testing-stage.html

Alfred Mann Foundation limb loss web pages at: http://aemf.org/item/limb-loss/

Battelle *New Device Allows Brain to Bypass Spinal Cord, Move Paralyzed Limbs* (Press

Release, no date). Available from: http://www.battelle.org/media/press-releases/new-device-allows-brain-to-bypass-spinal-cord-move-paralyzed-limbs

Blue Brain Project website at: http://bluebrain.epfl.ch/

BrainGate website at: http://braingate2.org

Cochlear website at: http://www.cochlear.com

Weio Gao et al 'Artificial Mircomotors in the Mouse's Stomach' *ACS Nano* (Volume 9 No 1, 2015). Available from: http://pubs.acs.org/doi/ipdf/10.1021/nn507097k

Gizmag 'Scientists Demonstrate A Mind-controlled Future' (4 November 2014). Available from: http://www.gizmag.com/go/3423/

Gizmag 'Braingate Brain-Machine-Interface Takes Shape' (6 December 2004). Available from: http://www.gizmag.com/go/3503/

Sarah Griffiths 'Student Creates Cyborg Cockroach That He Can Control With His THOUGHTS', *Daily Mail* (5 June 2015). Available from: http://www.dailymail.co.uk/sciencetech/article-3112404/Student-controls-cyborg-cockroach-MIND-Video-shows-insect-guided-maze-powered-thoughts.html

Human Brain Project FAQ. Available at: https://www.humanbrainproject.eu/en_GB/discover/the-project/faq

Michalle Ma 'UW Study Shows Direct Brain Interface Between Humans' *UW Today* (5 November 2014). Available from: http://www.washington.edu/news/2014/11/05/uw-study-shows-direct-brain-interface-between-humans/

Henry Markram et al 'Reconstruction and Simulation of Neocortical Microcircuitry', *Cell* (Volume 163 Issue 2, 8 October 2015). Abstract at: http://www.cell.com/cell/abstract/S0092-8674(15)01191-5

Bruze Mazlish *The Fourth Discontinuity: The Co-Evolution of Humans and Machines* (New Haven: Yale University Press, 1993).

Pierre Mégevand 'One more step along the long road towards brain-to-brain interfaces', *PLOS Blogs* (24 September 2015). Available from: http://blogs.plos.org/neuro/2015/09/24/one-more-step-along-the-long-road-towards-brain-to-brain-interfaces/

Myoelectric Prosthetics website at: http://www.myoelectricprosthetics.com/

Open Bionics website at: http://www.openbionics.com/

David Orenstein *People With Paralysis Control Robotic Arms Using Brain-computer Interface* (Brown University News Report, 16 May 2012). Available from: https://news.brown.edu/articles/2012/05/braingate2

Jacopo Prisco 'Will Nanotechnology Soon Allow You To "Swallow The Doctor"?' *CNN* (20 January 2015). Available from: http://edition.cnn.com/2015/01/29/tech/mci-nanobots-eth/

Antonio Regalado 'A Brain-Computer Interface That Works Wirelessly', *MIT Technology Review* (14 January 2015). Available from: http://www.technologyreview.com/news/534206/a-brain-computer-interface-that-works-wirelessly/

RT 'Uploading Human Brain For Eternal Life Is Possible – Cambridge Neuroscientist' (25 May 2015). Available from: https://www.rt.com/news/261909-brain-upload-computer-program/

Second Sight website at: http://www.secondsight.com

Stocco A, Prat CS, Losey DM, Cronin JA, Wu J, Abernethy JA et al. 'Playing 20 Questions with the Mind: Collaborative Problem Solving by Humans Using a Brain-to-Brain Interface', *PLoS ONE* (Volume 10, Issue 9, September 2015). Abstract available via: http://journals.plos.org/plosone/article?id=10.1371/journal.pone.0137303

Gregory Stock *Metaman: Humans, Machines and the Birth of a Global Super-organism* (London: Transworld Publishers, 1993).

Touch Bionics website at: http://www.touchbionics.com/

INDEX

23andMe, 252
2PP, 94, 95

3D printing, 3–5, 7, 15–42, 79–80, 91–4, 95, 102, 140, 150–2, 184–5, 209, 211, 215, 272–3, 275, 276, 294, 301, 302
3D Systems, 22, 209
3DXTech, 79

adverse drug response, 253, 255
aeroponics, 63
Affymetrix, 256
AGI, 5–6, 105–6, 127–9, 130, 131, 136
Agile Nanosat Platform, 210
AI, 5–6, 12, 97, 101–32, 184, 186, 239, 244, 249, 250, 257, 267, 269, 283, 290, 291, 294, 296, 300–1, 304
Airbus, 27–8, 183, 185
Aldebaran, 144–6, 150
Alfred E. Mann Foundation, 277
algae, 53, 62, 66
Alibaba, 146, 152
American Society for Testing and Materials, 17
Amyris, 54–5, 56, 61
Ancient Eqyptians, 303
Anderson, Eric C., 208
aneutronic fusion, 233–4, 235
APM, 4, 70, 74–5, 82–7, 89–90

Apollo Moon landings, 69, 72, 141, 166, 207, 213, 218, 225, 227, 231, 304
Apple, 5, 106, 107, 146, 243
AquaAdvantage salmon, 46, 263
Argus II retinal prosthesis, 274–5
Arianespace, 177, 229
Arkyd spacecraft, 209
Armstrong, Neil, 218, 239
Armstrong, Rachel, 65–6
Artemisinin Project, 56
Artifical consciousness, 136–7
artificial general intelligence, *see* AGI
artificial intelligence, *see* AI
ASIMO robot, 142–4
Asimov, Isaac, 156
assemblers, 87
Assurex, 254–5
Asterank.com, 198
asteroid bases, 203, 204
asteroid capture, 206–8
asteroid composition, 192–5
asteroid mining, 8, 176, 202–12, 288
Asteroid Redirect Mission, 200-202, 207, 210
ASTM, 17
Astrobotic, 217, 220, 227, 228
Atala, Anthony, 38
Atlas, 146–7
atomically precise manufacturing, *see* APM

attractors, 69, 140
Audi, 124, 140
augmented reality, 117
autonomous vehicles, 123–7, 141
autonomous vehicles, 6
Avery, Oswald, 45, 247

BAAM, 31-32
Baidu, 124, 140
Battelle, 277–8, 281
batteries, 76–7, 78
Big Data, 108-110, 111–12, 128, 129, 253, 257, 300
BigRep, 30–1
binder jetting, 22–4
BioBricks, 49
biofuels, 4, 54–5, 57, 60, 96
biological electronics, 50–51, 59, 61, 128–9, 266
biomodules, 286–7
bionic eyes, 273–5
bioplastics, 4, 53, 57, 60, 66–7, 96
bioprinting, 3, 9, 34–40, 43–4, 92–3, 121, 139, 157, 267, 273, 283, 288
Blair, Brad R., 202
Blue Brain Project, 289
Boeing, 177, 201, 219
bone printing, 273
Boston Consulting Group, 152–3
Boston Dynamics, 146–7
Boyer, Hebert, 46
Bozzato, Fabrizio, 236
brain interfaces, 278–83, 286
Brain, Marshall, 155–6
BrainGate, 281–2
Brown, Louise, 261
Brown, Spencer C., 174, 175
Brundtland Commission, 162
Burke, Martin D., 94, 95
Burkhart, Ian, 278

C-3PO, 133, 155

Calgene, 46
carbon nanotubes, 20, 76–7, 79, 86, 180
Carlson Curve, 256, 257
Carlson, Rob, 256
CCTV cameras, 116, 120
Celera Genomics, 247
Centre for Bionic Medicine, 276
Chandrayaan-1, 218
Chang'e 3, 218, 224
Chang'e Project, 223
Chinese Dream, 236
Chinese lunar Exploration Program, see CLEP
Chinese National Space Administration, 178, 217
Cincinnati Incorporated, 31
Clarke, Arthur C.,11
CLEP, 223–5
click-and-spit genetic tests, 252
climate change, 62, 66, 89, 167-8, 236
Clinton, Bill, 72
closed systems, 161–2, 188, 212
cochlear implants, 273
Cognitec, 118–19
cognitive computing, 107-8
Cohen, Stanley, 46
Colaprete, Anthony, 216
Collins, Francis, 248
ColorFabb, 20
concrete 3D printing, 21, 28–9
Constellation Program, 201, 218
convergence, 41–2, 91–3, 97, 128, 259, 294, 295, 301
Copernicus, Nicolaus, 270
Cortana, 5, 106, 107, 253, 300
COTS Program, 219, 221
Crick, Francis, 46, 247
Criswell, David, 237
Critchlow, Hannah, 291
CRON diet, 246
C-Type asteroids, 193, 196

Cyberkinetics, 282
cyborgs, 268–9, 271–96, 303
cystic fibrosis, 250, 252, 260, 262

Daimler, 124, 125, 140
DARPA Robotics Challenge, 147–8
DARPA, 45, 147–8, 300
Darwin, Charles, 270
data driven medicine, 113–15, 258
DDM, 26
De Lorenzo, Victor, 48–9
Deep Space Industries, 8, 209–11, 212
Delta-v, 191–2
depopulation, 188, 212, 238
designer babies, 261–3
Despommier, Dickson, 63
determinism, *see* technological determinism
Diamandis, Peter, 208
diminished reality, 117
direct digital manufacturing, 26
directed energy deposition, 22, 24–5
discontinuities, 270–2
distributed manufacturing facilities, 25, 42, 60
dolphins, 103–4
Dragon capsule, 177, 178, 185, 201, 222–3
DRC-HUBO robot, 147
Drexler, Eric, 69–71, 73–4, 75, 83, 86–7, 89–90, 163
driverless cars, *see* autonomous vehicles
drug carriers, 284–5
D-Wave Systems, 130

EBAM, 20–21
EEG, 279–81, 283
eighth continent, 216–17
Electron beam additive manufacturing, 20–21
electronic brain, 289–92, 294
Energia, 177, 225, 229
Engines of Creation, 69-71, 86
Enlitic, 113–14
enzymes, 84, 86–7
Eubanks, T.M., 230
European Space Agency, 199, 213, 215, 217, 225
Ex Machina, 137
ExOne, 23–4, 26
expansion ports (human), 286–7

Fab Labs, 33
Fabrisonic, 24
face payment systems, 120–21
face recognition, 118–21
Facebook, 106–7, 119–20, 253, 300
factories on legs, 60–1, 266
Falcon 9 rocket, 178, 185, 209, 223, 228
family balancing, 262
Fasotec, 30, 31
FDM, 18
Fertility Institutes, 262–3
filament, 18, 19–20, 79–80
Flavr Savr tomatoe, 46
foldamers, 85–6, 92
Ford, Martin, 154
Forgacs, Gabor, 34–5
Fourth Discontinuity, the, 270–2, 293
Foxconn, 146, 152
Freud, Sigmund, 270
Frey, Thomas, 69
Fukumuro, Yasuyuki, 181
fused deposition modelling, 18

Garagin, Yuri, 225
Garland, Alex, 137
Gates, Bill, 56, 101, 300
gene circuits, 50–1

gene doping, 265
gene expression microarrays, 255–6
gene therapy, 249, 258–61, 268
Genentech, 46, 263
Gene-RADAR, 250–1
General Motors, 134, 135
GeneSight, 254–5
GeneTherapy.net, 259
genetic testing, 249–53, 254–8, 262
GenoCAD, 50
genome sequencing, 115, 247–8, 255–8
genomic upgrading, 9, 249, 263–9
germline therapies, 264, 267
Glaser, Peter, 166, 167
GM crops, 46–7, 58, 263, 267
Google, 105, 106, 124, 125, 130, 140, 146, 152, 227, 243
Google Lunar XPRIZE, 217, 226–8
Google Now, 106, 253, 300
Google Translate, 121
Graphene 3D Lab, 20, 79–80
graphene, 20, 76–8, 79, 80, 86
Great Ormond Street Hospital, 260

Hackerspaces, 33
Haines, Agi, 287
Hawking, Steven, 101, 102, 131
helium-3, 8, 214, 232, 233–6, 237
Herceptin, 254
HiSeq X genome sequencer, 256–7
HIV viral load test, 251
Honda, 138-9, 141–4
Huawei, 122
Hull, Charles, 21–2
Human Brain Project, 289–90
human evolution, 9–10, 11, 88, 101, 102, 132, 238–40 246, 249, 262, 263, 268–9, 285–8, 293–6
Human Genome Project, 247–9, 256

Humanity 2.0, 292–6
Humanity+, 245–6
humanized organs, 60–1, 266–7, 273
humanoid robots, 6–7, 10, 12, 124, 128, 129, 133–157, 244, 266, 269, 272, 286, 294, 301
Huxley, Julian, 244

IAA, 168–73, 174, 176, 181
IBM, 76, 82-3, 105, 107-8, 110–11, 113, 130, 136, 289
iGEM, 49
Illumina, 256
iMakr, 29
immune system, 284
InMoov, 150–1
Intel, 74, 76
International Academy of Astronautics, *see* IAA
International Launch Services, 177, 229
International Space Station, *see* ISS
Internet of Things, 109, 117
Internet Revolution, 1, 295, 298, 299-300
Intrexon, 44–5, 55
invitro fertilization, *see* IVF
ion engines, 199–200, 207
ISS, 177, 184, 209, 214, 219
IVF, 261-63

J. Craig Venter Institute, *see* JCVI
Jacob, Francis, 47
Jade Rabbit, 224
Jain, Naveen, 228
JAXA, 8, 175, 177, 181–2, 198, 199, 217, 226
JCVI, 51–2
Jeopardy, 108

KAIST, 53, 147
Kaya, Nobuyuki, 175

Keck Institute for Space Studies, 206–8
Khoshnevis, Behrokh, 21
Kurzweil, Ray, 105

L'Oreal, 40
Langevin, Gael, 150, 151
language translation, 121–22, 279, 300
laser power transmission, 170–1, 183, 237
laser sintering, 24
LCROSS, 215–16, 218, 220
LDM, 2–5, 6–7, 9–10, 15, 41–2, 95–7, 102, 131, 139, 140, 184, 187, 211, 239, 244, 286, 288, 294, 302
Lewis, John S., 209, 212
Lewis, Randy, 59–60
Limits to Growth, The, 161–2, 163
limpet ships, 204, 205
Lin Industrial, 226
lipoplexes, 260
local digital manufacturing, *see* LDM
Local Motors, 31–2, 34, 60
localization, 30–32, 53, 97, 140–41
loposomes, 260
LSP, 237
Luna 2, 217
Luna 7, 218
Lunar CATALYST, 219–20, 228
Lunar Flashlight, 221
lunar mining, *see* mining the Moon
lunar solar power, *see* LSP
lunar space elevator, 229–30

Made in Space, 184
magic, 11, 12, 304
Main Asteroid Belt, 189–90, 198
Maisonnier, Bruno, 144
Maker Movement, 33, 140
Makies, 29

malaria, 56
Mankins, John, 165–66, 173, 174, 175, 185
Mari, Masahiro, 136
MarkForged, 20
mass customization, 29–30
Masten Space Systems, 217, 220
Mata, Alvaro, 85
material extrusion, 16–21, 28, 31
material jetting, 21, 22
Mazlish, Bruce, 270–1, 293
McKinsey Global Institute, 115
Mcor, 25
medical AI, *see* data driven medicine
mental automation, 101–2, 105–6
Merkle, Ralph, 87, 88
Metaman, 292–3
microfabricators, 4–5, 11, 25, 74–5, 89–90, 95–7, 106, 140, 184, 185, 198, 257, 294, 301, 302, 304
microfactories, 32, 33, 42, 60
MicroGravity Foundry, 211
microprocessors, 74, 75–6, 88, 95, 106, 128, 129
Microsoft, 5, 106, 107, 122, 130, 243, 300
microwave power transmission, 166, 167, 169, 172, 173, 174–6, 237
Miescher, Johan Friedrich, 247
mining the Moon, 176, 213–40
modular species, 288, 295
molecule-making machine, 94, 95
Monod, Jacques, 47
Monsanto, 46, 263
Moon bases, 218, 221–2, 225, 226, 237
Moon energy, 232–7
Moon Express, 216–17, 220, 227, 228
Moon Mail, 228
Moore, Gordon, 256

More, Max, 244–5
M-Type asteroids, 194
Musk, Elon, 101, 124, 131, 185
Myelectrics, 275–7

Nakamura, Makoto, 34
Nanobiosym, 250–1, 257
nanobots, 284–5
nanocomposites, 20, 79–82, 96
nanolithography, 75-76
Nanotechnology 2.0, 74–5, 83, 102, 301
nanotechnology, 4–5, 42, 69–97, 139, 140, 157, 185, 187, 250, 259, 267, 271, 283, 284, 294, 301
Nanotex, 81
NAO robot, 144–5, 150
narrative, 303–4
narrow AI, 105–6, 127–8
NASA, 8, 69, 130, 148, 149, 167, 168, 173, 176, 184, 199, 200–1, 206, 210, 215, 217, 218–21, 223, 225, 304
National Centre for Nanoscience & Technology, 72
National Nanotechnology Initiative, *see* NNI
Near-Earth asteroids, *see* NEAs,
NEAs, 190–1, 192, 193, 196, 197, 198, 200, 201, 206–7, 208
Nelson, Bradley, 285
Nervous System, 29
neural networks, 115–16, 130, 250
Neurobridge, 277–8
NexGen Space, 221–2, 232
NNI, 71–3
non-viral vectors, 260–1
Novogen MMX, 36
nuclear fission, 232
nuclear fusion, 214, 232–4, 235

Oak Ridge National Laboratory, 31–2

Open Bonics, 275, 276
Open Plant Initiative, 58
Orbital ATK, 177, 229
organic computing, 7, 50–1, 61, 62, 266
Organovo, 9, 35–7, 40
Orion capsule, 200–2, 218, 220
Outer Space Treaty, 231–2
Ozbolat, Ibrahim, 37-8, 283
Ozgene, 265

Part-Time Scientists, 227, 228
Pell, Barney, 228
Pepper robot, 145–6
PGD, 262
pharmacogenomics, 249, 253-5
pillars of civilization, 303–4
PLA, 19, 53, 80
Planetary Resources, 8, 208–9, 210
plant synthetic biology, 58–9
platinum, 8, 194, 197, 202, 204, 214
polylactic acid, *see* PLA
polyplexes, 260
Poppy Humanoid, 152
positional assembly, 82-3
post-genomic medicine, 247–69, 283, 288, 303
Potentially hazardous asteroids, 190–1
powder bed fusion, 22, 24
pre-implantation genetic diagnosis, *see* PGD
Printrbot, 18
production tooling, 26
Project for Emerging Nanotechnologies, 71
prosthetics, 271, 272–2
protein engineering, 71, 84–6, 89, 93, 96, 259
protocels, 65–6

Qiang, Yang, 122
Qualcom Tricorder XPRIZE, 251

quantum computing, 129–30

R5, 148, 149
radioresistance, 268
Ranger 7, 217
Registry of Standard Biological Parts, 49
Repoxygen, 264–5
Resource Prospector, 220–1
Richards, Bob, 228
Robots in Gastronomy, 21
robots, see humanoid robots
Romeo robot, 146
Roscosmos, 217, 225

Sasaki, Susumu, 181, 182
SB1.0, 47–8, 51
SBOL, 50
SBSP, 8, 164–186, 196, 303
Schroepfer, Mike, 119
Sciaky, 20
Scott Crump, 17-18
Second Law of Thermodynamics, 161, 162-3, 188
Second Sight, 274
selective reverse mutation, 258
Selene lunar impactor, 218
self-assembly, 4, 83–7, 92-3, 173–4, 185, 259, 288
self-cleaning glass, 80
SETI, 155
Shackleton Energy Company, 217, 228–9
sheet lamination, 25
Shelly Cobra, 31-32
Singularity loop, 131-2
Singularity, the, 11–12, 131–2, 137–8, 184, 186, 268, 292, 293, 304–5
Sino-Russian Moon base, 226
Siri, 5, 106, 253, 300
Skype Translator, 122
SLA, 22, 23

SMOs, 44, 58–61
Society of Manufacturing Engineers, 26
Softbank, 144–6, 152
solar power satellites, 166–174, 178, 181–6, 196, 211, 303
solar power, 78, 164–86
Solazyme, 55
somatic therapies, 264
Sonter, Mark, 202
Soyuz, 177, 214
Space Adventures, 213–14
Space Dream (Chinese), 236
space elevators, 8, 178–80, 184, 196, 229–30
Space Launch System, 177, 201, 218
space race, 217–18, 223, 238
Space Shuttle, 166, 167
space-based digital manufacturing, 184–5, 211
space-based solar power, see SBSP
SpaceIL, 217, 227–8
SpaceX, 101, 177, 178, 185, 201, 209, 219, 222–3, 228, 229
Spiber, 55–6, 59
spider goats, 59–60, 266
spider silk, 55-6, 59-60
SPS-ALPHA, 173–4, 185
Star Trek, 251
Star Wars, 133
stereolithography, 22
Stock, Gregory, 292–3, 295
Stratasys, 18–19, 22, 23, 27
Strati, 31, 32
S-Type asteroids, 193–4*
Sugimoto, Maki, 31
Surveyor lunar probes, 218
sustainability, 7, 162–4, 187, 237, 294, 305
synthetic biology, 4, 6, 42, 43–68, 83, 86, 91–93, 95, 102, 139, 140, 157, 185, 187, 259, 271, 272, 283, 286, 294, 301

Synthetic Genomics Inc, 52, 60–1
synthetically modified organism, see SMO
SyntheticBiology.org, 48

Tamny, John, 153–4
Targetted muscle reinnervation, see TMR
Technical University of Vienna, 93, 94
technological determinism, 298–301
Tesla Motors, 101, 124, 140
thermoplastics, 18–20, 79
TMR, 276–7
tooling, 26
toothbrushes, 67
Touch Bionics, 275–6
transcriptor, 50
transgenics, 46, 59–60, 265–6
transhumanism, 9, 12, 88, 138, 205, 240, 244–6, 265, 266, 267–9, 272, 283–8, 291, 303, 304–5
tricorder, 251, 257
Tsiolkovsky, Constantin, 164
Turing, Alan, 103, 104, 105
Turing Test, 103–5
two-photon polymerization, see 2PP

Ultimaker, 18
ultracapacitors, 77–8
uncanny valley, 136
United Nations Environment Programme, 187
United Nations Office for Outer Space Affairs, 232
universal translator, 122
uploading, 291–2

urban agriculture, 63–5

Valkyrie robot, 148, 149
vat photopolymerization, 21–2
vectors, 258–61, 267
Venice, 66
vertical farms, 4, 63–5
viral vectors, 259–60
Virgin Galactic, 178
virtual assistants, 5, 50–1, 106–7, 109, 117, 128, 129, 253, 300
Vision for US Space Exploration, 218
vision recognition, 116–21
Volvo Trucks, 26
voxeljet, 23
V-Type asteroids, 194

WAAM, 20
Wake Forest Institute, 38–9
Warwick, Kevin, 105
Watson, 108, 110–11, 113, 136
Watson, James, 46, 247
WinSun Technologies, 28–9
Wire and arc additive manufacturing, see WAAM
wireless power transmission, 166, 167, 169, 170, 171, 172, 173, 174–6, 236–7
Woerner, Johan-Deitrich, 213
Xiji, Wang, 183

X-SCID, 259–60, 261
X-Type asteroids, 194

Yutu rover, 224

Zinser, Michael, 153
Ziyuan, Ouyang, 225, 236

Printed in Great Britain
by Amazon.co.uk, Ltd.,
Marston Gate.